SpringerBriefs in Applied Sciences
and Technology

SpringerBriefs in Computational Intelligence

Series Editor

Janusz Kacprzyk, Systems Research Institute, Polish Academy of Sciences,
Warsaw, Poland

SpringerBriefs in Computational Intelligence are a series of slim high-quality publications encompassing the entire spectrum of Computational Intelligence. Featuring compact volumes of 50 to 125 pages (approximately 20,000-45,000 words), Briefs are shorter than a conventional book but longer than a journal article. Thus Briefs serve as timely, concise tools for students, researchers, and professionals.

More information about this subseries at http://www.springer.com/series/10618

Patricia Melin · Emanuel Ontiveros-Robles ·
Oscar Castillo

New Medical Diagnosis Models Based on Generalized Type-2 Fuzzy Logic

 Springer

Patricia Melin
Division of Graduate Studies
Tijuana Institute of Technology
Tijuana, Baja California, Mexico

Emanuel Ontiveros-Robles
Division of Graduate Studies
Tijuana Institute of Technology
Tijuana, Baja California, Mexico

Oscar Castillo
Division of Graduate Studies
Tijuana Institute of Technology
Tijuana, Baja California, Mexico

ISSN 2191-530X   ISSN 2191-5318 (electronic)
SpringerBriefs in Applied Sciences and Technology
ISSN 2625-3704   ISSN 2625-3712 (electronic)
SpringerBriefs in Computational Intelligence
ISBN 978-3-030-75096-1   ISBN 978-3-030-75097-8   (eBook)
https://doi.org/10.1007/978-3-030-75097-8

This Springer imprint is published by the registered company Springer Nature Switzerland AG
The registered company address is: Gewerbestrasse 11, 6330 Cham, Switzerland

Preface

Nowadays, the evolution of intelligent systems for decision making has been increased considerably, as these systems are getting more intelligent and can be helpful to experts in making better and faster decisions. One of the more important realms in decision making is the area of medical diagnosis, and many kinds of intelligence systems provide the expert good assistance to diagnosis; some of these methods are, for example artificial neural networks (can be very powerful to find tendencies) and support vector machines that avoid overfitting problems and statistical approaches (e.g., Bayesian). However, the present research is focused on one of the most relevant kinds of intelligent system, namely fuzzy systems. The main objective of the present book is the generation of fuzzy diagnosis systems that offer competitive classifiers to be applied in diagnosis systems. To generate these systems, we have proposed a methodology for the automatic design of classifiers and focused on generalized type-2 fuzzy logic because of its uncertainty handling that can provide us with the necessary robustness to be competitive with other kinds of methods. In addition, different alternatives to the uncertainty modeling, rules selection, and optimization have been explored. Besides, different experimental results are presented as evidence of the good results obtained when compared with respect to conventional approaches and literature references based on fuzzy logic.

This book aims to be a reference for scientists and engineers interested in designing decision-making assistants based on computer science and looking for an alternative with noise robustness and uncertainty modeling. We consider that this book can also be used to inspire hybridizations of different methods with the proposed approaches or in the design of specific proposal methods for complex problems.

In Chap. 2, we present a brief introduction for the most relevant concepts required for the development of the proposed approaches, for example fuzzy logic or swarm intelligence algorithms, for optimization.

Chapter 3 introduces the proposed methodologies and approaches for the design of generalized type-2 complex systems, the proposed alternatives for membership functions modeling and rules selection, for example, and considerations for the hybridizing with other powerful algorithms.

Chapter 4 contains the experimental results of four variations proposed documenting results for different setups and benchmark problems in order to evaluate the performance with the conventional approach.

Chapter 5 presents the analysis of the obtained results, a cross-validation comparison, a statistical test in order to get evidence of improvement, and finally a time-computation comparison.

Chapter 6 contains the conclusions of this work with criteria about which considerations can be used in order to design a system for a specific application. In addition, some considerations of performance and computational cost are offered.

We end this preface of the book by giving thanks to all the people who have helped or encouraged us during the writing of this book. First of all, we would like to thank Dr. Patricia Melin, for always supporting our work and for motivating us to write our research work. We would also like to thank our colleagues working in fuzzy logic, which are too many to mention each by their name. Of course, we need to thank our supporting agencies, CONACYT and TNM, in our country, for their help during this project. We have to thank our institution, Tijuana Institute of Technology, for always supporting our projects. Finally, we thank our respective families for their continuous support during the time that we spend in this project.

Tijuana, Mexico   Prof. Patricia Melin
February, 2021   Dr. Emanuel Ontiveros-Robles
                                                                              Prof. Oscar Castillo

Contents

Chapter 1
Introduction

Nowadays, one of the most important applications of intelligent systems are expert systems for decision making, these kinds of systems can provide experts with a tool for expert systems. There exist different kinds of systems based on artificial intelligence, for example, Artificial Neural Networks (ANN) [1–5] and variations of these, other examples are the Support Vector Machines [6–10], Decision Trees [11–15], and others. The main contribution of this work is the exploration and design of new methodologies and approaches for the generation of systems that allow an improvement in the performance of Fuzzy Classifiers applied in medical diagnosis specifically. The idea is using General Type-2 Fuzzy Logic as the core of these new approaches, and this because these kinds of systems have demonstrated to be better than Interval Type-2 Fuzzy Systems and Type-1 Fuzzy Systems [16, 17]. The uncertainty handling influences the performance of General Type-2 Fuzzy Logic, and this is most evident when the uncertainty is justified for example in [18] the generation of the Footprint of uncertainty is a relevant part of the paper because it effects in the performance of a Fuzzy Classifier is significant.

The proposed model includes different stages for the automatic design of diagnosis systems that can help a medical expert to make decisions in diagnosis. Exploring also alternatives for reducing the computational cost, for example, the Type-reduction [19], the high order α-planes aggregation [20], and other alternatives for the selection of the rules and optimization. The main idea is improving the classification of existing systems with the inclusion of the uncertainty in the models, an example of this can be illustrated in the proposed approach that includes the combination of General Type-2 Fuzzy Logic and the Support Vector Machines classifiers. The stages and details are organized as follows. Chapter 2 presents the background theory relevant to the development of the proposed approaches, Chap. 3 presents the proposed methodologies and systems, starting with the basic elements of the systems and finally the proposed alternatives to be tested in the experimentation. Chapter 4 presents the designed experiments in order to compare the proposed systems with respect to conventional approaches from theory and literature, on the other hand, the obtained results are

© The Author(s), under exclusive license to Springer Nature Switzerland AG 2021
P. Melin et al., *New Medical Diagnosis Models Based on Generalized Type-2 Fuzzy Logic*,
SpringerBriefs in Computational Intelligence,
https://doi.org/10.1007/978-3-030-75097-8_1

discussed in Chap. 5, and finally Chap. 6 contains the conclusions about the research work and future works.

References

1. X. Xu, D. Cao, Y. Zhou, J. Gao, Application of neural network algorithm in fault diagnosis of mechanical intelligence. Mech. Syst. Signal Process. **141**, 106625 (2020). https://doi.org/10. 1016/j.ymssp.2020.106625
2. J. Jiang, H. Wang, J. Xie, X. Guo, Y. Guan, Q. Yu, Medical knowledge embedding based on recursive neural network for multi-disease diagnosis. Artif. Intell. Med. **103**, 101772 (2020). https://doi.org/10.1016/j.artmed.2019.101772
3. W. Xuan, G. You, Detection and diagnosis of pancreatic tumor using deep learning-based hierarchical convolutional neural network on the internet of medical things platform. Future Gener. Comput. Syst. S0167739X20307779 (2020). https://doi.org/10.1016/j.future.2020.04.037
4. X. Liu, Y. Zhou, Z. Wang, Recognition and extraction of named entities in online medical diagnosis data based on a deep neural network. J. Vis. Commun. Image Represent. **60**, 1–15 (2019). https://doi.org/10.1016/j.jvcir.2019.02.001
5. J.S. Majeed Alneamy, Z.A. Hameed Alnaish, S.Z. Mohd Hashim, R.A. Hamed Alnaish, Utilizing hybrid functional fuzzy wavelet neural networks with a teaching learning-based optimization algorithm for medical disease diagnosis. Comput. Biol. Med. **112**, 103348 (2019). https://doi.org/10.1016/j.compbiomed.2019.103348
6. M. Wang, H. Chen, Chaotic multi-swarm whale optimizer boosted support vector machine for medical diagnosis. Appl. Soft Comput. **88**, 105946 (2020). https://doi.org/10.1016/j.asoc. 2019.105946
7. A. Kampouraki, D. Vassis, P. Belsis, C. Skourlas, e-Doctor: a web based support vector machine for automatic medical diagnosis. Proc. Soc. Behav. Sci. **73**, 467–474 (2013). https://doi.org/ 10.1016/j.sbspro.2013.02.078
8. D. Conforti, R. Guido, Kernel based support vector machine via semidefinite programming: application to medical diagnosis. Comput. Oper. Res. **37**(8), 1389–1394 (2010). https://doi. org/10.1016/j.cor.2009.02.018
9. A. Subasi, Medical decision support system for diagnosis of neuromuscular disorders using DWT and fuzzy support vector machines. Comput. Biol. Med. **42**(8), 806–815 (2012). https:// doi.org/10.1016/j.compbiomed.2012.06.004
10. A. Viloria, Y. Herazo-Beltran, D. Cabrera, O.B. Pineda, Diabetes diagnostic prediction using vector support machines. Proc. Comput. Sci. **170**, 376–381 (2020). https://doi.org/10.1016/j. procs.2020.03.065
11. M.M. Ghiasi, S. Zendehboudi, A.A. Mohsenipour, Decision tree-based diagnosis of coronary artery disease: CART model. Comput. Methods Progr. Biomed. **192**, 105400 (2020). https:// doi.org/10.1016/j.cmpb.2020.105400
12. S. Naganandhini, P. Shanmugavadivu, Effective diagnosis of Alzheimer's disease using modified decision tree classifier. Proc. Comput. Sci. **165**, 548–555 (2019). https://doi.org/10.1016/ j.procs.2020.01.049
13. L.O. Moraes, C.E. Pedreira, S. Barrena, A. Lopez, A. Orfao, A decision-tree approach for the differential diagnosis of chronic lymphoid leukemias and peripheral B-cell lymphomas. Comput. Methods Progr. Biomed. **178**, 85–90 (2019). https://doi.org/10.1016/j.cmpb.2019. 06.014
14. S. Itani, F. Lecron, P. Fortemps, A one-class classification decision tree based on kernel density estimation. Appl. Soft Comput. **91**, 106250 (2020). https://doi.org/10.1016/j.asoc.2020.106250
15. C.K. Madhusudana, H. Kumar, S. Narendranath, Fault diagnosis of face milling tool using decision tree and sound signal. Mater. Today Proc. **5**(5), 12035–12044 (2018). https://doi.org/ 10.1016/j.matpr.2018.02.178

16. E. Ontiveros-Robles, P. Melin, O. Castillo, Comparative analysis of noise robustness of type 2 fuzzy logic controllers. Kybernetika, 175–201. https://doi.org/10.14736/kyb-2018-1-0175

17. E. Ontiveros, P. Melin, O. Castillo, Comparative study of interval Type-2 and general Type-2 fuzzy systems in medical diagnosis. Inf. Sci. **525**, 37–53 (2020). https://doi.org/10.1016/j.ins.2020.03.059

18. M.A. Sanchez, O. Castillo, J.R. Castro, Method for measurement of uncertainty applied to the formation of interval type-2 fuzzy sets, in *Design of Intelligent Systems Based on Fuzzy Logic, Neural Networks and Nature-Inspired Optimization*, vol. 601, ed. by P. Melin, O. Castillo, J. Kacprzyk (Springer International Publishing, Cham, 2015), pp. 13–25

19. E. Ontiveros-Robles, P. Melin, O. Castillo, New methodology to approximate type-reduction based on a continuous root-finding Karnik Mendel algorithm. Algorithms **10**(3), 77 (2017). https://doi.org/10.3390/a10030077

20. E. Ontiveros, P. Melin, O. Castillo, High order α-planes integration: a new approach to computational cost reduction of general type-2 fuzzy systems. Eng. Appl. Artif. Intell. **74**, 186–197 (2018). https://doi.org/10.1016/j.engappai.2018.06.013

Chapter 2
Background and Theory

In this chapter the theoretical elements necessary to the development of the proposed approach and methodologies are introduced. The organization of the chapter contains an introduction to Fuzzy Logic, the Fuzzy C-Means algorithm, and finally, the principal methods used for classification in the literature.

2.1 Fuzzy Logic

This section contains the introduction to the Fuzzy Logic, a brief overview of its development since its creation, the theory about Interval Type-2 Fuzzy Inference Systems, and finally the concepts of General Type-2 Fuzzy Inference Systems.

2.2 Type-1 Fuzzy Inference Systems

Type-1 Fuzzy Logic was introduced in 1965 [1] and consist in an expansion of the traditional logic, where, in sets theory, the membership values are binary. However, Zadeh proposes that the membership values can be a continuous value between 0 and 1 called membership degree, the notation of a Fuzzy Set is expressed in (2.1).

$$A = \{\mu_A(x) | \forall x \in X\} \tag{2.1}$$

Based on this kind of fuzzy set exist two different kinds of Type-1 Fuzzy Inference Systems. The details of the two main kinds of T1 FIS are presented next.

© The Author(s), under exclusive license to Springer Nature Switzerland AG 2021
P. Melin et al., *New Medical Diagnosis Models Based on Generalized Type-2 Fuzzy Logic*,
SpringerBriefs in Computational Intelligence,
https://doi.org/10.1007/978-3-030-75097-8_2

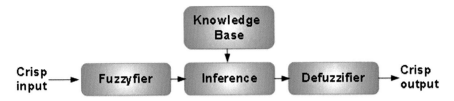

Fig. 2.1 Type-1 Mamdani FIS

2.3 Type-1 Mamdani Fuzzy Inference System

Type-1 Mamdani Fuzzy Inference System [2] is one of the most successfully applied in real-world applications [3–15], this is because it helps the computer to make decisions based on human knowledge.

The structure of this kind of system is composed of some process or steps that are illustrated in Fig. 2.1.

An explanation of this process is presented next.

Fuzzifier: This process realizes the conversion of the feature domain to the fuzzy domain obtaining the membership degree for the input data to its corresponding fuzzy set.

Rules Base and Fuzzy Inference Engine: This process consists of the evaluation of the fuzzy rules of the systems, these rules can be based on human knowledge or can be generated for some optimization algorithm. The inference is computed based on the *Modus ponens* inference as is expressed in (2.2).

$$R^l: IF \ x_1 \ is \ \widetilde{F}_1^l \ and \ldots and \ x_p \ is \ \widetilde{F}_p^l, \ THEN \ y \ is \ \widetilde{G}^l,$$
$$where \ l = 1, \ldots, M \tag{2.2}$$

Defuzzifier: Compute the calculation of the output of the system, this means, that convert the output fuzzy set in a crisp output to be applied for the real-world problem.

2.4 Type-1 Sugeno Fuzzy Inference System

Type-1 Sugeno Fuzzy Inference System [16] is the other kind of FIS most used for different kind of applications [17–22]. This kind of system is different from the Mamdani systems but the difference is only in one step of the computation.

The structure of this kind of system is composed of some process or steps that are illustrated in Fig. 2.2.

As can observed, these kinds of systems can be explained as Artificial Neural Networks with a specific architecture. An explanation of this process is presented next.

Fig. 2.2 Type-1 Sugeno FIS

Fuzzifier: Same that Mamdani systems, this process realizes the conversion of the feature domain to the fuzzy domain obtaining the membership degree for the input data to its corresponding fuzzy set.

Rules Base and Fuzzy Inference Engine: Same than Mamdani systems, this process consists of the evaluation of the fuzzy rules of the systems, these rules can be based on human knowledge or can be generated for some optimization algorithm. The inference is computed based on the *Modus ponens* inference as is expressed in (2.3).

$$R^l : IF \ x_1 \ is \ \tilde{F}_1^l \ and \ldots and \ x_p \ is \ \tilde{F}_p^l, \ THEN \ y \ is \ \tilde{G}^l,$$
$$where \ l = 1, \ldots, M \tag{2.3}$$

Output: The main difference between Sugeno and Mamdani Fuzzy Inference Systems is the defuzzifier because in Sugeno the fuzzy rules are associated with linear functions, and the output is the weighted average of every function and their rules firing force.

2.5   Interval Type-2 Fuzzy Inference Systems

On the other hand, Interval Type-2 Fuzzy Logic (IT2 FL) introduced by Mendel in [23, 24] is the evolution of Type-1 Fuzzy Logic that includes the uncertainty in its model. The main difference is in the concept of membership degree, this is because in the T1 FL the membership degree is a crisp value between 0 and 1, however, in IT2 FL the membership degree is an interval with two boundaries between 0 and 1. The mathematical expression of these systems can be found in (2.4).

$$A = \{\mu_A(x) | \forall x \in X\} \tag{2.4}$$

The main difference of IT2 FL with respect T1 FL can be observed in the membership functions. Considering that the membership degree in IT2 FL is an interval instead of a crisp value (T1 FL), the membership degree for systems based on IT2 FL can be modeled by two type-1 membership functions called *upper* and *lower*

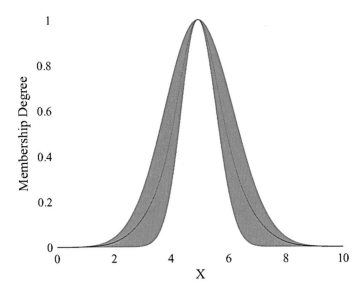

Fig. 2.3 IT2 MF example

membership functions. Figure 2.3 illustrates an example of the IT2 membership function.

The shadowed area in Fig. 2.3 delimited by the *upper* and *lower* MFs is called Footprint of Uncertainty (FOU) [24–26], this concept is directly related with the uncertainty and its mathematical representation is expressed in (2.5)

$$\text{FOU} \in \left[\underline{\mu}_{\tilde{A}}(x), \overline{\mu}_{\tilde{A}}(x)\right] \tag{2.5}$$

If the FOU is zero, the system based on IT2 FL is equivalent to a system based on T1 FL, so, the T1 FL is contained within IT2 FL.

On the other hand, in IT2 FL exist also different kinds of systems based on rules.

2.6   Interval Type-2 Mamdani Fuzzy Inference System

IT2 Mamdani FIS is probably the most IT2 FIS most used in the literature for a variety of problems, for example [27–33].

Have a structure very similar to their equivalent of T1 FL, but with some differences related to the uncertainty handling. This structure can be observed in Fig. 2.4.

An explanation of this process is presented next.

Fig. 2.4 IT2 Mamdani FIS

IT2 Fuzzifier: The process is the same as a T1 Mamdani FIS but considering the uncertainty handling. Then, IT2 Fuzzifier is composed of two T1 Fuzzifier, for the *upper* and the *lower* MFs.

IT2 Rules Base and Fuzzy Inference Engine: The same process as Type 1, but performing two times the same process, for *upper* and *lower* rules firing forces. The inference is computed based on the *Modus ponens* inference as is expressed in (2.6).

$$R^l: IF\ x_1\ is\ \tilde{F}_1^l\ and \ldots and\ x_p\ is\ \tilde{F}_p^l,\ THEN\ y\ is\ \tilde{G}^l,$$
$$where\ l = 1, \ldots, M \tag{2.6}$$

Type Reduction/Defuzzifier: Compute the calculation of the output of the system, this means, that convert the output fuzzy set in a crisp output to be applied for the real-world problem. This process is called *Type reduction* [34, 35], and represents the higher computational cost in this kind of system.

One of the most used type reduction methods is the called *Karnik–Mendel* [36] algorithm, the main idea is obtaining of the critical points of two functions that are combinations of the output upper and lower fuzzy set. These output functions are presented in (2.7) and (2.8) for the calculation of the left and right point respectively.

$$C_l = \operatorname*{min}_{\forall \theta_i \in \left[\overline{\mu}_{\tilde{A}}(z), \underline{\mu}_{\tilde{A}}(z)\right]} = \frac{\sum_{i=1}^{N} z_i \theta_i}{\sum_{i=1}^{N} \theta_i}$$
$$= \frac{\sum_{i=1}^{L} z_i \overline{\mu}_{\tilde{A}}(z_i) + \sum_{i=L+1}^{N} z_i \underline{\mu}_{\tilde{A}}(z_i)}{\sum_{i=1}^{L} \overline{\mu}_{\tilde{A}}(z_i) + \sum_{i=L+1}^{N} \underline{\mu}_{\tilde{A}}(z_i)} \tag{2.7}$$

$$C_r = \operatorname*{max}_{\forall \theta_i \in \left[\overline{\mu}_{\tilde{A}}(z), \underline{\mu}_{\tilde{A}}(z)\right]} = \frac{\sum_{i=1}^{N} z_i \theta_i}{\sum_{i=1}^{N} \theta_i}$$
$$= \frac{\sum_{i=1}^{R} z_i \underline{\mu}_{\tilde{A}}(z_i) + \sum_{i=R+1}^{N} z_i \overline{\mu}_{\tilde{A}}(z_i)}{\sum_{i=1}^{R} \underline{\mu}_{\tilde{A}}(z_i) + \sum_{i=R+1}^{N} \overline{\mu}_{\tilde{A}}(z_i)} \tag{2.8}$$

where z is the output domain, $\overline{\mu}(z)$ and $\mu(z)$ are the upper and lower MF of the output fuzzy set, and finally, L and R are the called *switch points*. The switch points are the

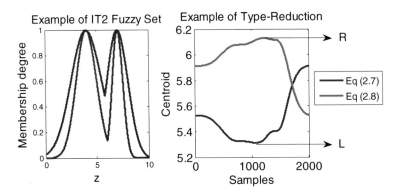

Fig. 2.5 Example type-reduction

points in the output domain where the combined functions centroids [presented in (2.7) and (2.8)] obtain their critical points (minimum and maximum respectively).

Figure 2.5 illustrates an example of these concepts. In the left can be observed an example of output Fuzzy Set and the main idea is the obtaining of the switch points. In the right plot, the Eqs. (2.7) and (2.8) are illustrated with their corresponding critics points which are the switch points in the presented example.

There exist different methodologies to obtain the switch points however Algorithm 2.1 details one of the most used in the literature, the Karnik–Mendel algorithm.

Algorithm 2.1 Karnik–Mendel Algorithm.

Step	Left point	Right point
1	Sort x_i by increasing order	Sort x_i by increasing order
2	Initialize w_i as $$wi = \frac{\overline{w_i} + \underline{w_i}}{2}$$	Initialize w_i as $$wi = \frac{w_i + \overline{w_i}}{2}$$
3	Compute $$y = \frac{\sum_{i=1}^{N} x_i w_i}{\sum_{i=1}^{N} w_i}$$	Compute $$y = \frac{\sum_{i=1}^{N} x_i w_i}{\sum_{i=1}^{N} w_i}$$
4	Find the switch point k where $x_k < y < x_{k+1}$	Find the switch point k where $x_k < y < x_{k+1}$
5	Set $$w_i = \begin{cases} \overline{w_i}, & i \le k \\ \underline{w_i}, & i > k \end{cases}$$	Set $$w_i = \begin{cases} \underline{w_i}, & i \le k \\ \overline{w_i}, & i > k \end{cases}$$
6	Compute $$y' = \frac{\sum_{i=1}^{N} x_i w_i}{\sum_{i=1}^{N} w_i}$$	Compute $$y' = \frac{\sum_{i=1}^{N} x_i w_i}{\sum_{i=1}^{N} w_i}$$
7	If $y == y'$ then stop, set $y_l = y$ and $L = k$, if not go to step 8	If $y == y'$ then stop, set $y_r = y$ and $R = k$ if not go to step 8

(continued)

(continued)

Step	Left point	Right point
8	Set $y = y'$ and go to step 3	Set $y = y'$ and go to step 3

As can noted, this process is iterative and requires a higher computational cost in comparison to the defuzzification of the T1 FIS, however, there exist many variations that allow the reduction of the computational cost as can be found in [34].

2.7   Interval Type-2 Sugeno Fuzzy Inference System

The other interesting kind of FIS is the IT2 Sugeno FIS, which is very similar to the T1 Sugeno FIS but with the uncertainty handling. This kind of system has been successfully used for complex problems for example [37–40].

The structure of this kind of system is composed of some process or steps that are illustrated in Fig. 2.6.

Similar to Type-1 FIS, this kind of IT2 FIS can be expressed as ANN, next are explained the layers of the architecture.

Fuzzifier: This process is the same as the explained for IT2 Mamdani FIS.

Rules Base and Fuzzy Inference Engine: The same inference than the explained for IT2 Mamdani FIS. The inference is computed based on the *Modus ponens* inference as is expressed in (2.6).

Output: In interval type-2 Sugeno FIS there not exist a single way for realizing the Type-reduction, for example in [41] are explained some alternatives, some of them are the KM reduction and the WM reduction.

2.8   General Type-2 Fuzzy Inference Systems

This kind of system, similar to Interval Type-2 Fuzzy Inference Systems, allows uncertainty handling. However, the principal difference between these approaches is the representation of the uncertainty in the model, for example: In the IT2 MFs,

Fig. 2.6   IT2 Sugeno FIS

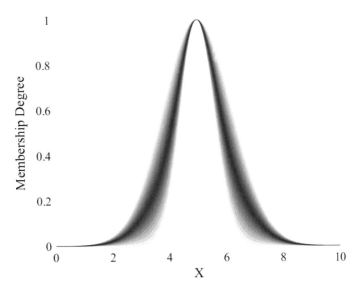

Fig. 2.7 GT2 MF

the fuzzy number (the result of the fuzzification) is an interval, on the other hand, in the GT2 MFs the fuzzy number is a T1 MF over a secondary axis that represents the uncertainty [42]. An illustration of a GT2 MF can be observed in Fig. 2.7.

As can be observed, the GT2 MF is, in fact, a collection of an infinite number of embedded T1 MFs over a secondary axis u that represents the uncertainty of the function.

The mathematical expression of a GT2 Fuzzy Set is presented in (2.9)

$$\tilde{A} = \left\{ \big((x, u), \mu_{\tilde{A}}(x, u)\big) \big| \forall u \in J_x \subseteq [0, 1] \right\} \qquad (2.9)$$

where $\mu_{\tilde{A}}(x, u)$ is the embedded T1 MF associated with the input domain (x) and the uncertainty domain (u).

If we observe this tridimensional function and consider only the boundaries of the function we can obtain the FOU of the GT2 MF, as is expressed in (2.10)

$$FOU\big(\tilde{A}\big) = \left\{ (x, u) \in \{X \times [0, 1]\} \big| \mu_{\tilde{A}}(x, u) > 0 \right\} \qquad (2.10)$$

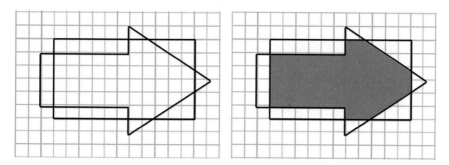

Fig. 2.8   Weiler–Atherton clipping example

2.9   Approximations of General Type-2 Fuzzy Inference Systems

As can be noted, the theoretical model of GT2 FIS is very difficult to compute because the computational cost of computing embedded Type-1 Fuzzy Sets is very high. However, exist some ways to approximate this model allowing the computation for real-world problems. There are presented the most used approaches for approximate the GT2 FISs.

2.10   Geometric Approximation

This approach was proposed by Coupland and John in [43] and consists on implement the concepts of computational geometry to the Fuzzy Logic (Fig. 2.8).

One of the backgrounds applied in this approach is the T-Norm as a clipping problem, in this case, solved by the Weiler-Atherton clipping [44]. Applying these concepts also for design one of the most original approaches of Type-2 Fuzzy Logic as can be observed in Fig. 2.9.

Some examples of this approach in the literature can be found in [45, 46].

2.11   α-planes Approximation

Introduced in [47], this approximation allows the computation of GT2 FISs for real-world applications. The main idea is the apply the concept of α-cut for every embedded T1 MF in the system. Applying this concept, we obtain horizontal slices called α-planes (2.11) that are equivalents to IT2 MFs.

$$\tilde{A}_\alpha = \left\{ (x, u) \in \{X \times [0, 1]\} | \mu_{\tilde{A}}(x, u) > \alpha \right\} \tag{2.11}$$

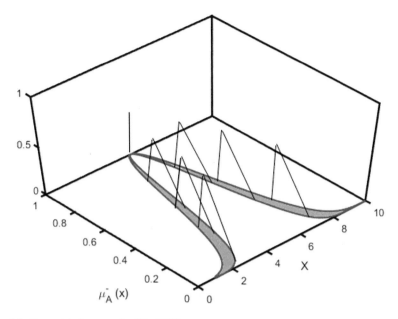

Fig. 2.9 Geometrical approach of the FOU

Based on this concept, we can discretize the GT2 MFs (tridimensional functions) with horizontal slices (equivalents to an IT2 MF) and solve every α-plane by separated to aggregate the results of every single α-plane, this aggregation is expressed in (2.12).

$$\widetilde{\widetilde{A}} = \bigcup \widetilde{A}_\alpha$$

$$\widetilde{\widetilde{A}} = \frac{\sum_{i=1}^{A} \alpha_i{}^{\alpha_i}\widetilde{A}_i(x')}{\sum_{i=1}^{A} \alpha_i} \tag{2.12}$$

where A is the number of α-planes, α_i is the ith α and ${}^{\alpha_i}\widetilde{A}_i(x')$ is the result of the ith α-plane. As can be observed, the aggregation of the α-planes is realized by a weighted average [48].

However, recently we propose an alternative to realizing the aggregation helped with the Newton–Cotes quadrature to reduce the number of α-planes required for a good approximation, the general form of Newton–Cotes is expressed in (2.13).

$$\int_a^b \left(\sum_{i=1}^{N-1} \left(\prod_{\substack{j=0 \\ j \neq i}}^{N-1} \frac{x - x_j}{x_i - x_j} \right) f(x) \right) dx \tag{2.13}$$

Consists in the integral of a Lagrange interpolator, applying this concept in the α-planes is possible to reduce the computational cost as is demonstrated in [49].

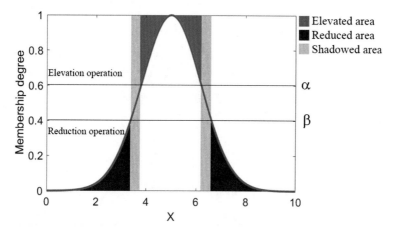

Fig. 2.10 Shadowed MF example

2.12 Shadowed Type-2 Fuzzy Inference Systems

This approximation is based on the concepts of Shadowed Fuzzy Sets [50–52] and the concepts of α-planes, the main idea consists in the use of the concept of Shadowed Fuzzy Sets in the embedded T1 MFs of the secondary axis.

The concept of Shadowed Fuzzy Sets consists of the reduction of the complexity of the T1 MFs based on optimized α-cuts. The main idea is the operation of reduction an elevation of two optimized areas trough two optimized values called α and β (Fig. 2.10).

The objective function of this optimization is expressed in (2.14)

$$V(\alpha, \beta) = \left| \int_{x \in A_r} \mu_A(x)dx + \int_{x \in A_e} (1 - \mu_A(x))dx - \int_{x \in S} dx \right| \qquad (2.14)$$

2.13 Type-2 Membership Functions

In this section are presented some Type-2 membership functions implemented in the proposed models, the main focus of these functions is the uncertainty model with mathematical expressions.

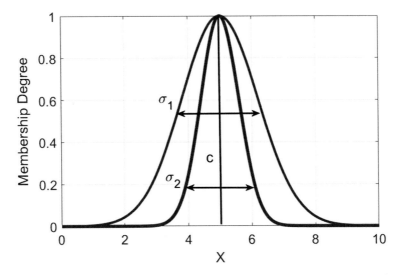

Fig. 2.11 Gaussian with uncertainty in sigma

2.14   Type-2 Gaussian Membership Functions

2.14.1   Gaussian MF with Uncertainty in the σ

The boundaries of the FOU for this GT2 MF are defined by an IT2 MF that represents a Gaussian MF with uncertainty in the σ. Figure 2.11 illustrates this function and (2.15) expresses the mathematical representation.

$$\sigma GaussIT2MF = \begin{cases} \overline{\mu}_t(x) = e^{-\frac{1}{2}\left(\frac{x-c}{\sigma_1}\right)^2} \\ \underline{\mu}_t(x) = e^{-\frac{1}{2}\left(\frac{x-c}{\sigma_2}\right)^2} \end{cases} \tag{2.15}$$

Based on the FOU (using Gauss IT2 MF) are proposed two variations of GT2 Gaussian MF with uncertainty in σ. The difference between both is the kind of secondary MF, using Triangular MF [41], and using Gaussian MF [41].

2.14.2   Gaussian MF with Uncertainty in the c

The boundaries of the FOU for this GT2 MF are defined by an IT2 MF that represents a Gaussian MF with uncertainty in the c. Figure 2.12 illustrates this function and (2.16) expresses the mathematical representation.

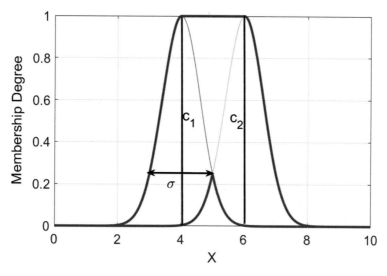

Fig. 2.12 Gaussian with uncertainty in c

$$cGauss\ IT2MF = \begin{cases} \overline{\mu}_t(x) = \begin{cases} e^{-\frac{1}{2}\left(\frac{x-c_1}{\sigma}\right)^2}, & x < c_1 \\ 1, & c_1 < x < c_2 \\ e^{-\frac{1}{2}\left(\frac{x-c_2}{\sigma}\right)^2}, & x > c_2 \end{cases} \\ \underline{\mu}_t(x) = \min\left(e^{-\frac{1}{2}\left(\frac{x-c_1}{\sigma}\right)^2}, e^{-\frac{1}{2}\left(\frac{x-c_2}{\sigma}\right)^2}\right) \end{cases} \qquad (2.16)$$

2.14.3 Double Gaussian MF with Uncertainty in the c and α

The boundaries of the FOU for this GT2 MF are defined by an IT2 MF that represents a Gaussian MF with uncertainty in the c. Figure 2.13 illustrates this function and (2.17) expresses the mathematical representation.

$$\mu_{\overline{A}}(x) = \begin{cases} \overline{\mu}_t(x) = \begin{cases} e^{\left(\frac{(-\frac{1}{2})(x-m_1)^2}{\sigma_1^2}\right)}, & x < m_1 \\ 1 & m_1 < x < m_2 \\ e^{\left(\frac{(-\frac{1}{2})(x-m_2)^2}{\sigma_1^2}\right)}, & x > m_2 \end{cases} \\ \underline{\mu}_t(x) \begin{cases} e^{\left(\frac{(-\frac{1}{2})(x-m_1)^2}{\sigma_2^2}\right)}, & x < \frac{(m_1+m_2)}{2} \\ e^{\left(\frac{(-\frac{1}{2})(x-m_2)^2}{\sigma_2^2}\right)}, & x > \frac{(m_1+m_2)}{2} \end{cases} \end{cases} \qquad (2.17)$$

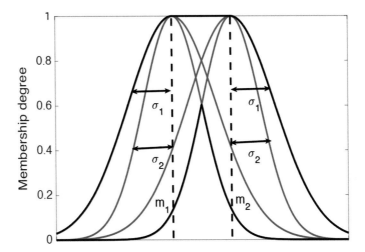

Fig. 2.13 Double Gaussian IT2 MF

2.14.4 Generalized Type-2 Membership Functions Forms

Based in the theory it is noticed that the Footprint of Uncertainty can be described by an Interval Type-2 membership function, on the other hand, if the FOU represents the boundaries of the uncertainty, can also be considered as the collection of the support of every secondary membership function. So, the counterpart, the collection of the core of every secondary membership function is called Core of Uncertainty (COU) and also be considered an IT2 membership function, based in this concepts the mathematical definition of a GT2 membership function with trapezoidal secondary membership function can be noticed in (2.18).

$$\widetilde{MF}_i(x, u) = \begin{cases} \frac{\underline{COU}(x) - u}{\underline{COU}(x) - \underline{FOU}(x)}, & \underline{FOU}(x) < u < \underline{COU}(x) \\ 1, & \underline{COU}(x) < u < \overline{COU}(x) \\ \frac{\overline{FOU}(x) - u}{\overline{FOU}(x) - \overline{CoU}(x)}, & \overline{COU}(x) < u < \overline{FOU}(x) \\ 0, & Otherwise \end{cases} \qquad (2.18)$$

where the FOU and the COU are IT2 membership functions, as can be note, if the upper and lower membership function of the COU are equals, the secondary membership function is a triangular membership function.

Fig. 2.14 Support vector
machine

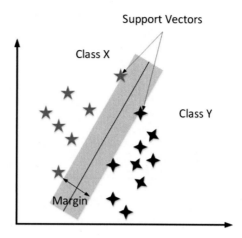

2.15 Classification Methods

This section introduces the theory about three of the most applied methods for classification and two strategies commonly used to improve the performance of the classifiers and its relation with the proposed approaches.

2.16 Support Vector Machine

This is one of the most effective tools in machine learning [53–57], a graphical example of this is illustrated in Fig. 2.14.

On the other hand, this method allows the implementation of the strategy of increasing the dimensionality of the data to perform the better separation of the classes, this process is obtained by the use of special functions called Kernels (Fig. 2.15).

In the proposed approaches, the concepts of hyperplane and kernel are used in a hybridization with Generalized Type-2 Fuzzy Classifiers obtaining a better performance that both approaches hybridized.

2.17 Decision Trees

A Decision Tree is an approach for decision support widely used in the computer sciences because providing robust systems that are interpretable and with a good classification rate [58–62]. Another advantage of these kinds of systems is that demands a low computational effort (Fig. 2.16).

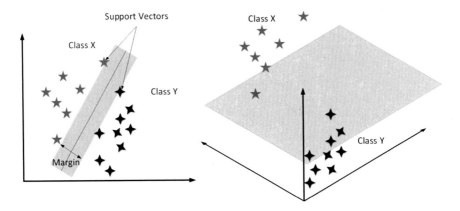

Fig. 2.15 Kernel example

Fig. 2.16 Decision tree
example—computational
effort of fuzzy systems

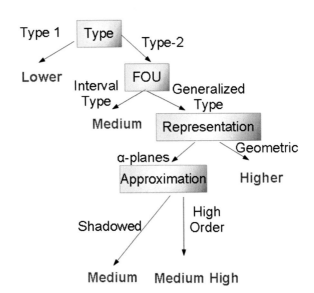

However, exist many kinds of Decision Trees a that is different based on the
methodology for training or pruning, Table 2.1 summarizes some of the most applied
kind of Decision Trees.

As can be observed exists many variations of DTs, in the literature can be found
in applications in different real problems for example in [63–67]. In the proposed
approach was explored the methodologies of C4.5 for the generations of the rules in
order to be applied in the generation of knowledge of fuzzy classifiers.

Table 2.1 Classification of decision trees

Kind	Used for	Scenario
Standard classification three	Classify	For two variables
C4.5	Classify/regression	For many variables
Standard regression three	Regression	For two variables
CART	Classify/regression	Two variables and for bagging approaches
CHAID	Classify	For categorical variables/multiple variables
QUEST	Classify/regression	For categorical and continuous variables/two variables

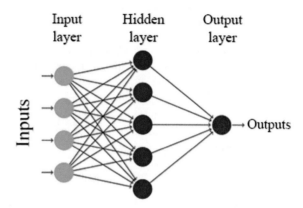

Fig. 2.17 Monolithic ANN

2.18 Artificial Neural Networks

Artificial Neural Networks (ANN) are inspires based on the human brain, which is a group of nodes interconnected that are organized in layers. Are one of the most powerful tools in the realm of artificial intelligence and are very versatile for be applied in many kinds of problems as can be noted in the literature [68–72]. Figure 2.17 illustrates an example of ANN.

There exist many variations of ANN-based on their architecture or the method for being trained, in Fig. 2.18 is illustrated a general view about the kind of ANN.

The ANNs have many hybridizations with other models, for example, in this work we present models based on the hybrid system Adaptive Neuro-Fuzzy Inference System (ANFIS) [73].

2.19 Bagging and Boosting

These methodologies for improving the performance of classifiers are based on the concepts of strong classifiers and weak classifiers. They are ensemble methods

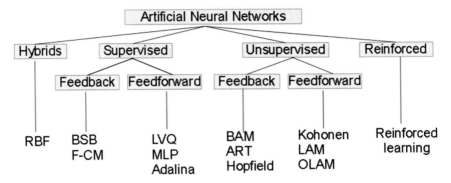

Fig. 2.18 Classification of ANNs

for decision-making problems, widely used in machine learning. The main idea of these approaches is the design of strong classifiers troughs weak classifiers. Remembering that many real-life problems can be viewed as classification problems. These methodologies are applied in order to improve the classification of the proposed approach, however, the most relevant is the inspiration of these methodologies for the uncertainty modelling proposed in the proposed classifiers, specifically, the uniform random sampling with replacement.

2.20 Bagging

Starting with the Bagging approach, the main idea is the collaboration of many models trained for different subsets of samples, this resampling is realized by a uniform random sampling with replacement of the original training set of samples, some examples of this approach in the literature can be found in [74–78]. Figure 2.19 illustrates the general concept of Bagging architecture.

The aggregation of the final output can be performed in different ways, but the main idea is that the use of these models we can obtain a better performance in comparison to a single model provided by the training of the original training set of samples.

2.21 Boosting

The Boosting approach is similar in its philosophy to the Bagging approach, the main idea is obtaining a strong classifier based on multiple weak classifiers, some examples of this approach in the literature can be found in [79–83]. The classifying process is very similar to the Bagging architecture but in the training, step is found the difference. The main idea of this approach is illustrated in Fig. 2.20.

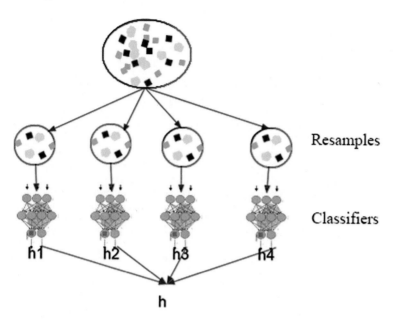

Fig. 2.19 Bagging architecture

Fig. 2.20 Boosting
architecture

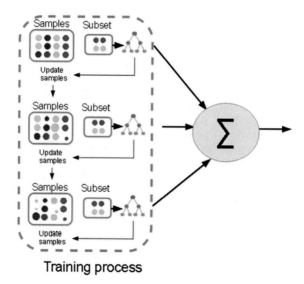

The main idea in this approach is the updating of the training set based on the performance of the previous classifier, this means, that the second classifier will be an expert in the data which the first classifier fails, and the third classifiers also will be more specialized. Every classifier has a weight and based on the classification and the weight has obtained the output.

References

1. L.A. Zadeh, Fuzzy sets. Inf. Control **8**(3), 338–353 (1965). https://doi.org/10.1016/S0019-995 8(65)90241-X
2. E.H. Mamdani, Application of fuzzy algorithms for control of simple dynamic plant. Proc. Inst. Electr. Eng. **121**(12), 1585 (1974). https://doi.org/10.1049/piee.1974.0328
3. E. Ontiveros-Robles, P. Melin, O. Castillo, Comparative analysis of noise robustness of type 2 fuzzy logic controllers. Kybernetika, 175–201 (2018). https://doi.org/10.14736/kyb-2018-1-0175
4. Y. Bai, D. Wang, On the comparison of type 1 and interval type 2 fuzzy logic controllers used in a laser tracking system. IFAC-Pap. **51**(11), 1548–1553 (2018). https://doi.org/10.1016/j.ifa col.2018.08.276
5. F. Cuevas, O. Castillo, Design and implementation of a fuzzy path optimization system for omnidirectional autonomous mobile robot control in real-time, in *Fuzzy Logic Augmentation of Neural and Optimization Algorithms: Theoretical Aspects and Real Applications*, vol. 749, ed. by O. Castillo, P. Melin, J. Kacprzyk (Springer International Publishing, Cham, 2018), pp. 241–252
6. R.-E. Precup, S. Preitl, E.M. Petriu, J.K. Tar, M.L. Tomescu, C. Pozna, Generic two-degree-of-freedom linear and fuzzy controllers for integral processes. J. Frankl. Inst. **346**(10), 980–1003 (2009). https://doi.org/10.1016/j.jfranklin.2009.03.006
7. M. Dhimish, V. Holmes, B. Mehrdadi, M. Dales, Comparing Mamdani Sugeno fuzzy logic and RBF ANN network for PV fault detection. Renew. Energy **117**, 257–274 (2018). https://doi.org/10.1016/j.renene.2017.10.066
8. Z. Ding, Z. Li, A cascade fuzzy control system for inverted pendulum based on Mamdani-Sugeno type, in *2014 9th IEEE Conference on Industrial Electronics and Applications*, Hangzhou, China, Jun 2014, pp. 792–797. https://doi.org/10.1109/ICIEA.2014.6931270
9. H. Pan, J. Yan, Study of Mamdani fuzzy controller and its realization on PLC, in *2006 6th World Congress on Intelligent Control and Automation*, Dalian, China, 2006, pp. 3997–4001.https://doi.org/10.1109/WCICA.2006.1713123
10. H. Ying, Conditions for general Mamdani fuzzy controllers to be nonlinear, in *2002 Annual Meeting of the North American Fuzzy Information Processing Society Proceedings. NAFIPS-FLINT 2002 (Cat. No. 02TH8622)*, New Orleans, LA, USA, 2002, pp. 201–203. https://doi.org/10.1109/NAFIPS.2002.1018055
11. J. Yan, H. Pan, Application of real-time FC based on geometrical meaning of Mamdani reasoning in air-cushioned headbox control system, in *2008 7th World Congress on Intelligent Control and Automation*, Chongqing, China, 2008, pp. 1218–1222. https://doi.org/10.1109/WCICA.2008.4594445
12. M. Mahfouf, M. Jamei, D.A. Linkens, Rule-base generation via symbiotic evolution for a Mamdani-type fuzzy control system, in *10th IEEE International Conference on Fuzzy Systems. (Cat. No. 01CH37297)*, Melbourne, Vic., Australia, 2001, vol. 1, pp. 396–399. https://doi.org/10.1109/FUZZ.2001.1007332
13. L. Mastacan, C.-C. Dosoftei, Temperature fuzzy control system with Mamdani controller in *2018 International Conference and Exposition on Electrical and Power Engineering (EPE)*, IASI, Oct 2018, pp. 0352–0356. https://doi.org/10.1109/ICEPE.2018.8559861.
14. P. Tarigan, S. Sinurat, M. Sinambela, Implementation of a Mamdani fuzzy logic controller for building automation using electronic control based on AT89S51, in *2015 International Conference on Technology, Informatics, Management, Engineering and Environment (TIME-E)*, Samosir, Toba Lake, Indonesia, Sept 2015, pp. 87–92. https://doi.org/10.1109/TIME-E.2015.7389753
15. M. Xu, H. Yu, L. Huang, G. Meng, The research of parallel structure and composite fuzzy control for linear machine based on Mamdani method, in *2014 17th International Conference on Electrical Machines and Systems (ICEMS)*, Hangzhou, China, Oct 2014, pp. 1749–1752. https://doi.org/10.1109/ICEMS.2014.7013777

16. T. Takagi, M. Sugeno, in *Fuzzy Identification of Systems and its Applications to Modeling and Control*, eds. by D. Dubois, H. Prade, R. R. Yager. Readings in fuzzy sets for intelligent systems (Morgan Kaufmann, 1993), pp. 387–403

17. A. Bemani-N, M.-R. Akbarzadeh-T, A hybrid adaptive granular approach to Takagi–Sugeno–Kang fuzzy rule discovery. Appl. Soft Comput. **81**, 105491 (2019). https://doi.org/10.1016/j.asoc.2019.105491

18. X. Gu, F.-L. Chung, S. Wang, Bayesian Takagi–Sugeno–Kang fuzzy classifier. IEEE Trans. Fuzzy Syst. **25**(6), 1655–1671 (2017). https://doi.org/10.1109/TFUZZ.2016.2617377

19. D. Krokavec, A. Filasová, A unitary construction of Takagi-Sugeno fuzzy fault detection filters. IFAC-Pap. **51**(24), 1193–1198 (2018). https://doi.org/10.1016/j.ifacol.2018.09.700

20. M. Shokouhifar, A. Jalali, Optimized Sugeno fuzzy clustering algorithm for wireless sensor networks. Eng. Appl. Artif. Intell. **60**, 16–25 (2017). https://doi.org/10.1016/j.engappai.2017.01.007

21. S.-H. Tsai, Y.-W. Chen, A novel identification method for Takagi-Sugeno fuzzy model. Fuzzy Sets Syst. **338**, 117–135 (2018). https://doi.org/10.1016/j.fss.2017.10.012

22. A. Zoukit, H. El Ferouali, I. Salhi, S. Doubabi, N. Abdenouri, Takagi Sugeno fuzzy modeling applied to an indirect solar dryer operated in both natural and forced convection. Renew. Energy **133**, 849–860 (2019). https://doi.org/10.1016/j.renene.2018.10.082

23. Q. Liang, J.M. Mendel, Interval type-2 fuzzy logic systems: theory and design. IEEE Trans. Fuzzy Syst. **8**(5), 535–550 (2000). https://doi.org/10.1109/91.873577

24. J.M. Mendel, R.I. John, F. Liu, Interval type-2 fuzzy logic systems made simple. IEEE Trans. Fuzzy Syst. **14**(6), 808–821 (2006). https://doi.org/10.1109/TFUZZ.2006.879986

25. H. Mo, F.-Y. Wang, M. Zhou, R. Li, Z. Xiao, Footprint of uncertainty for type-2 fuzzy sets. Inf. Sci. **272**, 96–110 (2014). https://doi.org/10.1016/j.ins.2014.02.092

26. J.E. Moreno et al., Design of an interval type-2 fuzzy model with justifiable uncertainty. Inf. Sci. (2019). https://doi.org/10.1016/j.ins.2019.10.042

27. Z. Du, Z. Yan, Z. Zhao, Interval type-2 fuzzy tracking control for nonlinear systems via sampled-data controller. Fuzzy Sets Syst. **356**, 92–112 (2019). https://doi.org/10.1016/j.fss.2018.02.013

28. M. El-Bardini, A.M. El-Nagar, Interval type-2 fuzzy PID controller for uncertain nonlinear inverted pendulum system. ISA Trans. **53**(3), 732–743 (2014). https://doi.org/10.1016/j.isatra.2014.02.007

29. A.M. El-Nagar, M. El-Bardini, Hardware-in-the-loop simulation of interval type-2 fuzzy PD controller for uncertain nonlinear system using low cost microcontroller. Appl. Math. Model. **40**(3), 2346–2355 (2016). https://doi.org/10.1016/j.apm.2015.09.005

30. P.J. Gaidhane, M.J. Nigam, A. Kumar, P.M. Pradhan, Design of interval type-2 fuzzy precompensated PID controller applied to two-DOF robotic manipulator with variable payload. ISA Trans. **89**, 169–185 (2019). https://doi.org/10.1016/j.isatra.2018.12.030

31. D.K. Jana, R. Ghosh, Novel interval type-2 fuzzy logic controller for improving risk assessment model of cyber security. J. Inf. Secur. Appl. **40**, 173–182 (2018). https://doi.org/10.1016/j.jisa.2018.04.002

32. R. Mohammadikia, M. Aliasghary, Design of an interval type-2 fractional order fuzzy controller for a tractor active suspension system. Comput. Electron. Agric. **167**, 105049 (2019). https://doi.org/10.1016/j.compag.2019.105049

33. J.R. Nayak, B. Shaw, B.K. Sahu, Application of adaptive-SOS (ASOS) algorithm based interval type-2 fuzzy-PID controller with derivative filter for automatic generation control of an interconnected power system. Eng. Sci. Technol. Int. J. **21**(3), 465–485 (2018). https://doi.org/10.1016/j.jestch.2018.03.010

34. M. Nie, W.W. Tan, Towards an efficient type-reduction method for interval type-2 fuzzy logic systems, in *2008 IEEE International Conference on Fuzzy Systems (IEEE World Congress on Computational Intelligence)*, Hong Kong, China, Jun 2008, pp. 1425–1432. https://doi.org/10.1109/FUZZY.2008.4630559

35. X. Liu, J.M. Mendel, D. Wu, Study on enhanced Karnik-Mendel algorithms: initialization explanations and computation improvements. Inf. Sci. **184**(1), 75–91 (2012). https://doi.org/10.1016/j.ins.2011.07.042

36. N.N. Karnik, J.M. Mendel, Centroid of a type-2 fuzzy set. Inf. Sci. **132**(1), 195–220 (2001). https://doi.org/10.1016/S0020-0255(01)00069-X
37. H. Li, S. Yin, Y. Pan, H.-K. Lam, Model reduction for interval type-2 Takagi-Sugeno fuzzy systems. Automatica **61**, 308–314 (2015). https://doi.org/10.1016/j.automatica.2015.08.020
38. G.M. Méndez, M. de los Angeles Hernández, Hybrid learning mechanism for interval A2-C1 type-2 non-singleton type-2 Takagi–Sugeno–Kang fuzzy logic systems. Inf. Sci. **220**, 149–169 (2013). https://doi.org/10.1016/j.ins.2012.01.024
39. M.A. Sanchez, O. Castillo, J.R. Castro, Information granule formation via the concept of uncertainty-based information with interval type-2 fuzzy sets representation and Takagi–Sugeno–Kang consequents optimized with Cuckoo search. Appl. Soft Comput. **27**, 602–609 (2015). https://doi.org/10.1016/j.asoc.2014.05.036
40. C. Martínez García, V. Puig, C.M. Astorga-Zaragoza, G.L. Osorio-Gordillo, Robust fault estimation based on interval Takagi-Sugeno unknown input observer. IFAC-Pap. **51**(24), 508–514 (2018). https://doi.org/10.1016/j.ifacol.2018.09.624
41. J.M. Mendel, *Uncertain Rule-Based Fuzzy Systems: Introduction and New Directions*, 2nd edn. (Springer International Publishing, Cham, 2017).
42. L.A. Lucas, T.M. Centeno, M.R. Delgado, General type-2 fuzzy inference systems: analysis, design and computational aspects, in *2007 IEEE International Fuzzy Systems Conference*, London, UK, Jun 2007, pp. 1–6. https://doi.org/10.1109/FUZZY.2007.4295522
43. S. Coupland, Type-2 fuzzy sets: geometric defuzzification and type-reduction, in *2007 IEEE Symposium on Foundations of Computational Intelligence*, Honolulu, HI, USA, Apr 2007, pp. 622–629. https://doi.org/10.1109/FOCI.2007.371537
44. E. Horowitz, M. Papa, Polygon clipping: analysis and experiences, in *Theoretical Studies in Computer Science* (Elsevier, 1992), pp. 315–339
45. S. Coupland, R. John, Geometric type-1 and type-2 fuzzy logic systems. IEEE Trans. Fuzzy Syst. **15**(1), 3–15 (2007). https://doi.org/10.1109/TFUZZ.2006.889764
46. S. Coupland, R. John, A fast geometric method for defuzzification of type-2 fuzzy sets. IEEE Trans. Fuzzy Syst. **16**(4), 929–941 (2008). https://doi.org/10.1109/TFUZZ.2008.924345
47. M. Mendel, F. Liu, D. Zhai, α-plane representation for type-2 fuzzy sets: theory and applications. IEEE Trans. Fuzzy Syst. **17**(5), 1189–1207 (2009).https://doi.org/10.1109/TFUZZ.2009.2024411
48. P. Melin, C.I. Gonzalez, J.R. Castro, O. Mendoza, O. Castillo, Edge-detection method for image processing based on generalized type-2 fuzzy logic. IEEE Trans. Fuzzy Syst. **22**(6), 1515–1525 (2014). https://doi.org/10.1109/TFUZZ.2013.2297159
49. E. Ontiveros, P. Melin, O. Castillo, High optimize α-planes integration: a new approach to computational cost reduction of general type-2 fuzzy systems. Eng. Appl. Artif. Intell. **74**, 186–197 (2018). https://doi.org/10.1016/j.engappai.2018.06.013
50. E. Ontiveros-Robles, P. Melin, A hybrid design of shadowed type-2 fuzzy inference systems applied in diagnosis problems. Eng. Appl. Artif. Intell. **86**, 43–55 (2019). https://doi.org/10.1016/j.engappai.2019.08.017
51. P. Melin, E. Ontiveros-Robles, C.I. Gonzalez, J.R. Castro, O. Castillo, An approach for parameterized shadowed type-2 fuzzy membership functions applied in control applications. Soft Comput. **23**(11), 3887–3901 (2019). https://doi.org/10.1007/s00500-018-3503-4
52. O. Castillo et al., Shadowed type-2 fuzzy systems for dynamic parameter adaptation in harmony search and differential evolution algorithms. Algorithms **12**(1), 17 (2019). https://doi.org/10.3390/a12010017
53. W.-J. Chen, Y.-H. Shao, C.-N. Li, M.-Z. Liu, Z. Wang, N.-Y. Deng, ν-projection twin support vector machine for pattern classification. Neurocomputing **376**, 10–24 (2020). https://doi.org/10.1016/j.neucom.2019.09.069
54. W.H. Asquith, The use of support vectors from support vector machines for hydrometeorologic monitoring network analyses. J. Hydrol. **583**, 124522 (2020). https://doi.org/10.1016/j.jhydrol.2019.124522
55. B. Richhariya, M. Tanveer, A reduced universum twin support vector machine for class imbalance learning. Pattern Recognit. 107150 (2020). https://doi.org/10.1016/j.patcog.2019.107150

56. M. Wadkar, F. Di Troia, M. Stamp, Detecting malware evolution using support vector machines. Expert Syst. Appl. **143**, 113022 (2020). https://doi.org/10.1016/j.eswa.2019.113022

57. S. Weerasinghe, S.M. Erfani, T. Alpcan, C. Leckie, Support vector machines resilient against training data integrity attacks. Pattern Recognit. **96**, 106985 (2019). https://doi.org/10.1016/j.patcog.2019.106985

58. M.M. Ghiasi, S. Zendehboudi, A.A. Mohsenipour, Decision tree-based diagnosis of coronary artery disease: CART model. Comput. Methods Progr. Biomed. **192**, 105400 (2020). https://doi.org/10.1016/j.cmpb.2020.105400

59. S. Naganandhini, P. Shanmugavadivu, Effective diagnosis of Alzheimer's disease using modified decision tree classifier. Proc. Comput. Sci. **165**, 548–555 (2019). https://doi.org/10.1016/j.procs.2020.01.049

60. L.O. Moraes, C.E. Pedreira, S. Barrena, A. Lopez, A. Orfao, A decision-tree approach for the differential diagnosis of chronic lymphoid Leukemias and peripheral B-cell lymphomas. Comput. Methods Programs Biomed. **178**, 85–90 (2019). https://doi.org/10.1016/j.cmpb.2019.06.014

61. S. Itani, F. Lecron, P. Fortemps, A one-class classification decision tree based on kernel density estimation. Appl. Soft Comput. **91**, 106250 (2020). https://doi.org/10.1016/j.asoc.2020.106250

62. C.K. Madhusudana, H. Kumar, S. Narendranath, Fault diagnosis of face milling tool using decision tree and sound signal. Mater. Today Proc. **5**(5), 12035–12044 (2018). https://doi.org/10.1016/j.matpr.2018.02.178

63. M. Gohari, A.M. Eydi, Modelling of shaft unbalance: modelling a multi discs rotor using K-nearest neighbor and decision tree algorithms. Measurement **151**, 107253 (2020). https://doi.org/10.1016/j.measurement.2019.107253

64. D. Hao et al., Application of decision tree in determining the importance of surface electro-hysterography signal characteristics for recognizing uterine contractions. Biocybern. Biomed. Eng. **39**(3), 806–813 (2019). https://doi.org/10.1016/j.bbe.2019.06.008

65. T. Lan, Y. Zhang, C. Jiang, G. Yang, Z. Zhao, Automatic identification of spread f using decision trees. J. Atmospheric Sol.-Terr. Phys. **179**, 389–395 (2018). https://doi.org/10.1016/j.jastp.2018.09.007

66. A. Trabelsi, Z. Elouedi, E. Lefevre, Decision tree classifiers for evidential attribute values and class labels. Fuzzy Sets Syst. **366**, 46–62 (2019). https://doi.org/10.1016/j.fss.2018.11.006

67. S.-B. Yang, T.-L. Chen, Uncertain decision tree for bank marketing classification. J. Comput. Appl. Math. **371**, 112710 (2020). https://doi.org/10.1016/j.cam.2020.112710

68. B. Cortés, R. Tapia Sánchez, J.J. Flores, Characterization of a polycrystalline photovoltaic cell using artificial neural networks. Sol. Energy **196**, 157–167 (2020). https://doi.org/10.1016/j.solener.2019.12.012

69. Zh.A. Dayev, Application of artificial neural networks instead of the orifice plate discharge coefficient. Flow Meas. Instrum. **71**, 101674 (2020). https://doi.org/10.1016/j.flowmeasinst.2019.101674

70. M. Ghazvini, H. Maddah, R. Peymanfar, M.H. Ahmadi, R. Kumar, Experimental evaluation and artificial neural network modeling of thermal conductivity of water based nanofluid containing magnetic copper nanoparticles. Phys. Stat. Mech. Appl. 124127 (2020). https://doi.org/10.1016/j.physa.2019.124127

71. S.H. Kim, S.G. Shin, S. Han, M.H. Kim, C.H. Pyeon, Feasibility study on application of an artificial neural network for automatic design of a reactor core at the Kyoto University Critical Assembly. Prog. Nucl. Energy **119**, 103183 (2020). https://doi.org/10.1016/j.pnucene.2019.103183

72. A.H. Zaji, H. Bonakdari, H.Z. Khameneh, S.R. Khodashenas, Application of optimized artificial and radial basis neural networks by using modified genetic algorithm on discharge coefficient prediction of modified labyrinth side weir with two and four cycles. Measurement **152**, 107291 (2020). https://doi.org/10.1016/j.measurement.2019.107291

73. J.-S.R. Jang, ANFIS: adaptive-network-based fuzzy inference system. IEEE Trans. Syst. Man Cybern. **23**(3), 665–685 (1993). https://doi.org/10.1109/21.256541

74. S. Agarwal, C.R. Chowdary, A-Stacking and A-Bagging: Adaptive versions of ensemble learning algorithms for spoof fingerprint detection. Expert Syst. Appl. **146**, 113160 (2020). https://doi.org/10.1016/j.eswa.2019.113160

75. J. Lin, H. Chen, S. Li, Y. Liu, X. Li, B. Yu, Accurate prediction of potential druggable proteins based on genetic algorithm and Bagging-SVM ensemble classifier. Artif. Intell. Med. **98**, 35–47 (2019). https://doi.org/10.1016/j.artmed.2019.07.005

76. S. Moral-García, C.J. Mantas, J.G. Castellano, M.D. Benítez, J. Abellán, Bagging of credal decision trees for imprecise classification. Expert Syst. Appl. **141**, 112944 (2020). https://doi.org/10.1016/j.eswa.2019.112944

77. S.E. Roshan, S. Asadi, Improvement of Bagging performance for classification of imbalanced datasets using evolutionary multi-objective optimization. Eng. Appl. Artif. Intell. **87**, 103319 (2020). https://doi.org/10.1016/j.engappai.2019.103319

78. Z. Wu et al., Using an ensemble machine learning methodology-Bagging to predict occupants' thermal comfort in buildings. Energy Build. **173**, 117–127 (2018). https://doi.org/10.1016/j.enbuild.2018.05.031

79. N. García-Pedrajas, Supervised projection approach for boosting classifiers. Pattern Recognit. **42**(9), 1742–1760 (2009). https://doi.org/10.1016/j.patcog.2008.12.023

80. A.A. Naem, N.I. Ghali, A.A. Saleh, Antlion optimization and boosting classifier for spam email detection. Future Comput. Inform. J. **3**(2), 436–442 (2018). https://doi.org/10.1016/j.fcij.2018.11.006

81. E. Volna, M. Kotyrba, Enhanced ensemble-based classifier with boosting for pattern recognition. Appl. Math. Comput. **310**, 1–14 (2017). https://doi.org/10.1016/j.amc.2017.04.019

82. J.J. Rodríguez, J. Maudes, Boosting recombined weak classifiers. Pattern Recognit. Lett. **29**(8), 1049–1059 (2008). https://doi.org/10.1016/j.patrec.2007.06.019

83. Y. Xu, D. Wang, W. Zhang, P. Ping, L. Feng, Detection of ventricular tachycardia and fibrillation using adaptive variational mode decomposition and boosted-CART classifier. Biomed. Signal Process. Control **39**, 219–229 (2018). https://doi.org/10.1016/j.bspc.2017.07.031

Chapter 3
Proposed Methodology

The present chapter presents the proposed methodology for the design of General Type-2 Fuzzy Inference Systems for diagnosis problems. The methodology presented is the core of the proposed variations of General Type-2 Fuzzy Inference Systems, it is composed of three general steps as can be observed in Fig. 3.1. The proposed approach of GT2 FIS will consider a Sugeno GT2 FIS.

As can observed, the methodology consists in only three steps but many variations of GT2 FIS can be designed by changing different ways for example for obtaining the fuzzy sets parameters, or the rules selection, or obtaining of the Sugeno parameters, the objective of the methodology is that can be used as a base for the development of better approaches for diagnosis problems.

The architecture of the proposed GT2 FIS is very similar to the original Adaptive Neuro-Fuzzy Inference System proposed by Jang in 1993 [1], but with the difference of be a GT2-A0, this system had been used as a base for the proposed variations. The proposed architecture can be found in Fig. 3.2.

Considering the presented architecture, next are introduced the steps of the methodology proposed.

3.1 Fuzzy Sets Parameters Obtaining

The parameters of the antecedent Fuzzy Sets can be obtained based on the data, as we propose, in the Fuzzy C-Means algorithm [2]. Other methods had been proved for the clusters obtaining, for example, K-means, subtractive, but the best results have been obtained with the FCM algorithm.

The main idea is the use of this clustering algorithm to obtain the clusters of the data and a membership degree of every data to every cluster, based on these membership degrees will be estimated the parameters of the input MFs.

P. Melin et al., *New Medical Diagnosis Models Based on Generalized Type-2 Fuzzy Logic*, SpringerBriefs in Computational Intelligence, https://doi.org/10.1007/978-3-030-75097-8_3

Fig. 3.1 Methodology
proposed

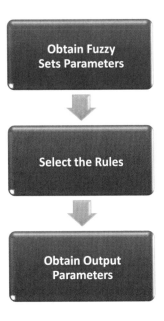

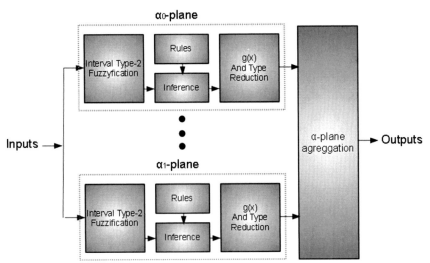

Fig. 3.2 Architecture of proposed GT2 FIS

An example of the clustering and the membership degrees can be observed in Fig. 3.3.

For example, the selected number of clusters for the FCM algorithm is two, and next is presented the most relevant approaches for the modeling of the membership functions proposed in the present research project.

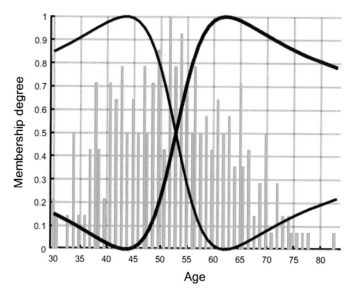

Fig. 3.3 Age attribute of the Haberman's dataset

3.2 Statistical Measures to Obtain the Parameters of the MFs

This first approach is the simplest way proposed and is based on statistical measures to obtain the parameters of the MFs (Algorithm 3.1).

Algorithm 3.1 Statistical approach [3]

Step	Action
1	Realize a uniform random sampling with replacement obtaining M subsets
2	Compute the fuzzy C-means algorithm for every subset obtaining M sets of N clusters with their respective center and membership degrees
3	Estimate the membership function (T1 Gaussian MF) parameters of every cluster and every subset ($M \cdot N$ Gaussians functions). Where c of the Gaussian is equivalent to the center of the cluster (obtained with the FCM algorithm) and the σ is the standard deviation of the membership degree of the corresponding cluster
4	Estimate the parameters of the GT2 GaussS() based on the set of T1 Gaussians MFs $c = mean(C_i)$; $\sigma_1 = \min(S_i)$; $\sigma_2 = \max(S_i)$ where C_i is the collection of the M (on per every subset) cs corresponding to the ith MF (of the N clusters) and S is the collection of the M (on per every subset) σs corresponding to the ith MF (of the N clusters)

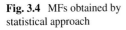
Fig. 3.4 MFs obtained by statistical approach

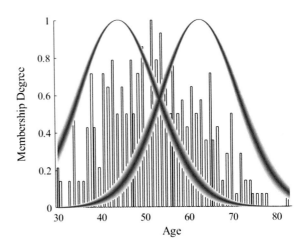

The final result of the use of this approach can be observed in Fig. 3.4.

As can observed, the uncertainty obtained is symmetric and compared with the distribution of the data appreciated in the normalized histogram, which is good modeling of the clusters.

3.3 Gaussian Regression

This approach is based on the principles of non-linear regression, the idea is obtaining statistically the best value of σ for every Gaussian function based on the center provided for the Fuzzy C-means algorithm, we propose this approach in [3, 4]. The methodology is presented in Algorithm 3.2.

Algorithm 3.2 Gaussian Regression approach

Step	Action
1	Realize a uniform random sampling with replacement obtaining M subsets
2	Compute the fuzzy C-means algorithm for every subset obtaining M sets of N clusters with their respective center and membership degrees
3	Estimate the membership function (T1 Gaussian MF) parameters of every cluster and every subset ($M \cdot N$ Gaussians functions). Where c of the Gaussian is equivalent to the center of the cluster (obtained with the FCM algorithm) and $$\sigma = \sqrt{\frac{\sum_{i=1}^{n}\left(\mu(x_i)-c\right)^4}{-2(\ln|y_i|)\sum_{i=1}^{n}\left(\mu(x_i)-c\right)^2}}$$ where $\mu(x_i)$ is the membership degree provided for the Fuzzy C-means algorithm and c is the center of the Gaussian

(continued)

Fig. 3.5 MFs obtained by
Gaussian regression

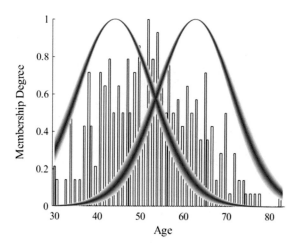

(continued)

Step	Action
4	Estimate the parameters of the GT2 GaussS() based on the set of T1 Gaussians MFs $c = mean(C_i)$; $\sigma_1 = \min(S_i)$; $\sigma_2 = \max(S_i)$ where C_i is the collection of the M (on per every subset) cs corresponding to the ith MF (of the N clusters) and S is the collection of the M (on per every subset) σs corresponding to the ith MF (of the N clusters)

The final result of the use of this approach can be observed in Fig. 3.5.

The obtained MFs also have symmetric uncertainty and based on the regression they are the optimum Gaussian MFs.

3.4 Optimized with Metaheuristics

The present alternative to obtaining of the Type-2 MFs parameters is one of the most demanding approaches because is based on the optimization of the FOU of every MF based on metaheuristics. For the present work, we only apply the Particle Swarm Optimization method. The methodology is summarized in Algorithm 3.3.

Algorithm 3.3 Gaussian Regression approach

Step	Action
1	Compute the fuzzy C-means algorithm for every subset obtaining M sets of N clusters with their respective center and membership degrees
2	Estimate the membership function (T1 Gaussian MF) parameters of every cluster and every subset ($M \cdot N$ Gaussians functions). Where c of the Gaussian is equivalent to the center of the cluster (obtained with the FCM algorithm) and $$\sigma = \sqrt{\frac{\sum_{i=1}^{n}\left(\mu(x_i)-c\right)^4}{-2(\ln|y_i|)\sum_{i=1}^{n}\left(\mu(x_i)-c\right)^2}}$$ where $\mu(x_i)$ is the membership degree provided for the Fuzzy C-means algorithm and c is the center of the Gaussian
3	Set the system rules
4	Estimate the output Sugeno coefficients (Type-1) based on the LSE method
4	Based on the complete model that optimizes the footprint of uncertainty of every input MF. This can be computed with your better metaheuristic optimization method

The final result of the use of this approach can be observed in Fig. 3.6. As can be observed, this methodology for the modelling of the Type-2 MFs offers a FOU optimized based in the data and this can be considered a justified uncertainty, but this can variate depending on the metaheuristic selected and the stochastic property of the metaheuristics.

Interestingly, the MFs obtained with metaheuristics because have not an interpretation but in many cases, they consider higher uncertainty and perform good results.

Fig. 3.6 MFs obtained optimizing with PSO

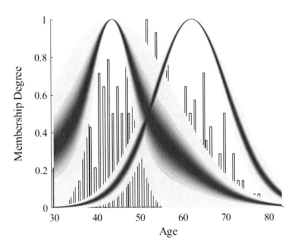

3.5 Asymmetric Uncertainty

The following methodology for the T2 MFs parameters obtaining is the most computational expensive, this is because requires additional computations for example concepts of boxplot and metaheuristics (Algorithm 3.4). However, this methodology provides us with asymmetric uncertainty that can be considered as a better modelling of uncertainty in the systems.

Algorithm 3.4 Asymmetric uncertainty approach

Step	Action
1	Realize a uniform random sampling with replacement obtaining M subsets
2	Compute the fuzzy C-means algorithm for every subset obtaining M sets of N clusters with their respective center and membership degrees
3	Estimate the membership function (T1 Gaussian MF) parameters of every cluster and every subset ($M \cdot N$ Gaussians functions). Where c of the Gaussian is equivalent to the center of the cluster (obtained with the FCM algorithm) and $$\sigma = \sqrt{\frac{\sum_{i=1}^{n}\left(\mu(x_i)-c\right)^4}{-2(\ln|y_i|)\sum_{i=1}^{n}\left(\mu(x_i)-c\right)^2}}$$ where $\mu(x_i)$ is the membership degree provided for the Fuzzy C-means algorithm and c is the center of the Gaussian
4	Obtain the FOU (Footprint of Uncertainty) and COU (Core of Uncertainty) helped with the Quartiles of every point in the domain of every feature FOU = [Min, Max], COU = [Q1, Q3] The result of this computation is a collection of data that can be plotted as two Interval Type-2 MFs
5	Based on the obtained FOU and COU find the best IT2 MF that describe the FOU and the COU, this can be performed helped with metaheuristics
6	Finish

The final result of the use of this approach can be observed in Fig. 3.7.

The use of embedded Type-1, boxplot and metaheuristics for a generation the GT2 MFS is very interesting because it provides us of asymmetric uncertainty that is not usual in the real world applications, the main idea is obtaining of a justified uncertainty that model by a better form the data uncertainty. However, is important to see how this uncertainty modelling impact in the performance of the systems.

3.6 Rules Selection

In the present section are explained some different ways to computing the rules selection.

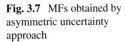

Fig. 3.7 MFs obtained by asymmetric uncertainty approach

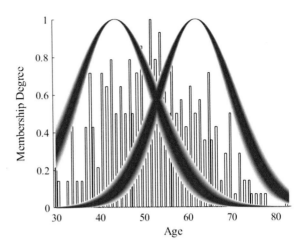

3.7 Based on ANFIS Architecture

The original architecture proposed by Jang in his first model of Adaptive Neuro-Fuzzy Inference System (ANFIS), where the number of rules are determined for the number of clusters and is predefined by the architecture, is not a full connected architecture and this allows the scalability of this architecture for many attributes or clusters (Fig. 3.8).

These simple rules provide us with systems that are very good classifiers, the main advantage is that these systems are scalable and they are very robust.

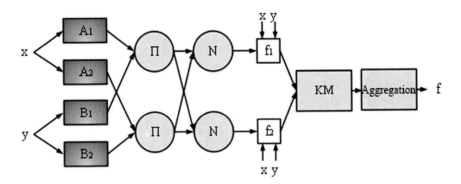

Fig. 3.8 ANFIS architecture

Fig. 3.9 Descendent gradient

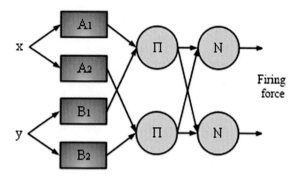

3.8 Descendent Gradient

The main idea in this approach is the training of an ANN equivalent to the first two stages of a Fuzzy Inference System, the fuzzification, and the inference/rules.

This architecture considers the output of the ANN as the linear combination of the firing force of the fuzzy rules in a fully connected set of rules (ignoring the Sugeno polynomial). The idea is that the bias of the third layer (firing forces) starts with the value of 0 (rules not connected) and after the optimization finalize with a value that will be normalized. The reason for optimizing the proposed architecture is because we can determine how much a rule participates in the decision of the fuzzy diagnosis and eliminate or aggregate the relevant rules. The ANN architecture proposed in this approach is illustrated in Fig. 3.9.

The algorithm proposed for this optimization is summarized in Algorithm 3.5 as follows.

Algorithm 3.5 Descendent Gradient Rules Generation

Step	Action
1	Compute the fuzzy C-means algorithm for every subset obtaining M sets of N clusters with their respective center and membership degrees
2	Estimate the membership function (T1 Gaussian MF) parameters of every cluster and every feature ($M \cdot N$ Gaussians functions). Where c of the Gaussian is equivalent to the center of the cluster (obtained with the FCM algorithm) and $$\sigma = \sqrt{\frac{\sum_{i=1}^{n} (\mu(x_i)-c)^4}{-2(\ln|y_i|)\sum_{i=1}^{n} (\mu(x_i)-c)^2}}$$ where $\mu(x_i)$ is the membership degree provided for the Fuzzy C-means algorithm and c is the center of the Gaussian
3	Set the bias on 0 (every connection with the output layer is a rule)
4	Compute the methodology of Descendent Gradient to optimize the connection with the output (to find how many influences every rule in the output)
5	Set the bias in 0 if (value < β) and 1 if (value > β) where β is a threshold (default value is 0.5) obtained by experimentation

(continued)

(continued)

Step	Action
6	If the bias does not change, then finish Else compute step 4

This way for rules generation provide us with the relevant rules and this helps to obtain very specifics classifications, however, there are susceptible to the overfitting problem.

3.9 Fuzzy Entropy

The Fuzzy Entropy [5] which is inspired in Shannon Entropy [6] can be applied in the generation of the fuzzy rules of the proposed classifiers. The concept is applied to select which rules have the most relevance in the system, this parameter can be computed as the fuzzy entropy of the firing forces and the classification, the rules which have the lower Fuzzy Entropy are the rules which more information contains and consequently, the better rules. A representation of Fuzzy Entropy can be observed in Fig. 3.10.

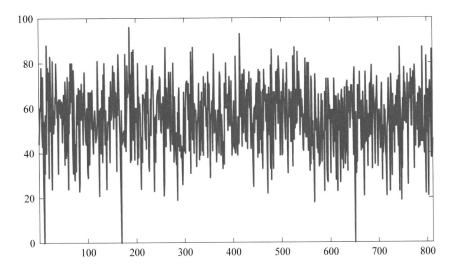

Fig. 3.10 Fuzzy entropy rules generation

The algorithm for this approach is summarized in Algorithm 3.6.

Algorithm 3.6 Fuzzy Entropy rules generation

Step	Action
1	Compute the fuzzy C-means algorithm for every subset obtaining M sets of N clusters with their respective center and membership degrees
2	Estimate the membership function (T1 Gaussian MF) parameters of every cluster and every feature ($M \cdot N$ Gaussians functions). Where c of the Gaussian is equivalent to the center of the cluster (obtained with the FCM algorithm) and $$\sigma = \sqrt{\frac{\sum_{i=1}^{n}(\mu(x_i)-c)^4}{-2(\ln|y_i|)\sum_{i=1}^{n}(\mu(x_i)-c)^2}}$$ where $\mu(x_i)$ is the membership degree provided for the Fuzzy C-means algorithm and c is the center of the Gaussian
3	Compute every possible rule obtaining the corresponding firing forces
4	Compute the fuzzy entropy of every rule obtaining the parameter δ
5	Set the δ in 0 if (value $< \beta$) and 1 if (value $> \beta$) where β is a threshold (default value is 0.5) obtained by experimentation
6	Finish

In this methodology the rules generated are similar to the obtained by a Decision three and the main idea is obtaining systems with relevant rules, however, for classification, the systems obtained are very susceptible to the overfitting and this affects their performance.

3.10 Sugeno Parameters Obtaining

The parameters of the Sugeno polynomial (Sugeno coefficients) are very relevant in the performance of the diagnosis system and need to be well estimated, in this work we propose to estimate these coefficients based on (Least Square Errors) LSE algorithm (Fig. 3.11).

The methodology for the LSE method consists of the minimization of the classification error modifying the Sugeno coefficients, this can be performed by partial derivate and finding the minimum point. An expression of this can be found in (3.1)

$$\left\{ \frac{\partial S_r}{\partial c_0^1}, \frac{\partial S_r}{\partial c_1^1}, \ldots, \frac{\partial S_r}{\partial c_m^1}, \ldots, \frac{\partial S_r}{\partial c_0^n}, \frac{\partial S_r}{\partial c_1^n}, \ldots, \frac{\partial S_r}{\partial c_m^n} \right\} = 0 \qquad (3.1)$$

where c_0^n is the zero Sugeno coefficients of the nth rule. In this step, the obtained Sugeno coefficients are the same for every α-plane.

Fig. 3.11 Least square error

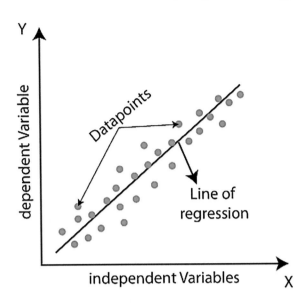

3.11 General Type-2 Fuzzy Inference Systems for Diagnosis Problems

Based on the concepts introduced in Chap. 1 and the methodology introduced in Chap. 2, this chapter presents an introduction to the different variations of GT2 FIS designed for diagnosis problems there exist many variations of GT2 FIS that can be designed based on hybridization or combination of the introduced concepts and algorithms however we present the most successfully GT2 FIS developed in this work.

Table 3.1 introduces the variations proposed and after will be explained in depth.

Every variation of GT2 FIS designed for diagnosis problems is explained in-depth as follows.

Table 3.1 Better approaches obtained

Name	Inputs	MFs	Antecedent parameters estimation	Rules selection	Consequent parameters estimation
GT2	Singleton	Dgauss	Embedded type 1	1 versus 1	LSE
ST2 PSO	Singleton	Sgauss	Type-1 + PSO	1 versus 1	LSE
NS GT2	No singleton	Dgauss	Embedded type-1	1 versus 1	LSE
SVM GT2	Singleton	Sgauss	Embedded type-1	1 versus 1	LSE
ASY GT2	Singleton	Mixed	Embedded type-1 + PSO + Boxplot	1 versus 1	LSE

3.12   General Type-2 Fuzzy Diagnosis System (GT2 FDS)

The General Type-2 Fuzzy Diagnosis System is the first approach presented in this work, consists of a singleton system with double Gaussian MFs with uncertainty in the c and the σ, as is explained in Table 3.1. Figure 3.12 illustrates the flowchart for the design of this system, in addition an example is presented for a better understanding.

For the example, is considered the example of Haberman's dataset for the present example, as can noted, we consider two attributes called "years old" and "operation year" in Fig. 3.13 can be observed in the histogram of both of the attributes.

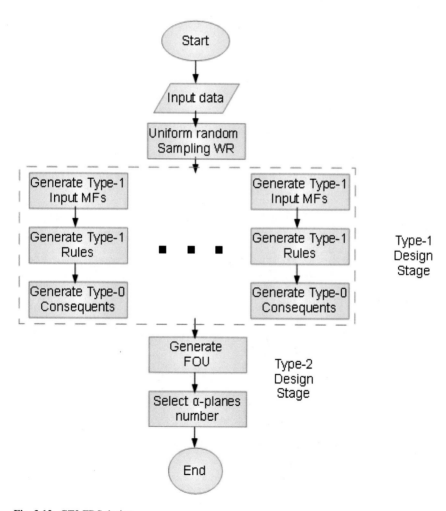

Fig. 3.12 GT2 FDS design

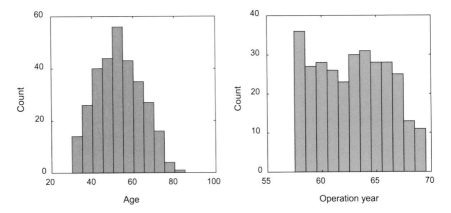

Fig. 3.13 Attributes histogram

Following the methodology proposed in 2.12 by using the principle of embedded T1 MFs for design the GT2 MFs. Figure 3.14 illustrates the multiple T1 MFs and Fig. 3.15 illustrates the generated GT2 MFs.

This system considers the fixed rules proposed by Jang in the original architecture of ANFIS (Sect. 2.21) and obtaining the rules illustrated in Fig. 3.16.

Finally, the Sugeno coefficients are obtained by the LSE algorithm detailed in Sect. 2.3.

The classification map (the surface provided by the classifier for every combination of input data) can be observed in Fig. 3.17.

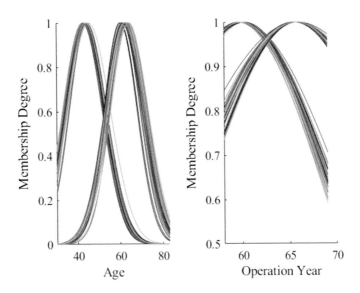

Fig. 3.14 Embedded type-1 MFs

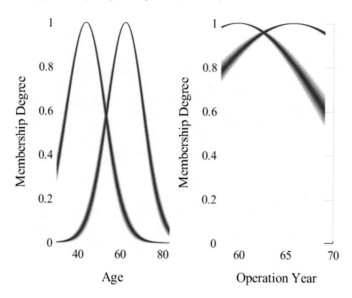

Fig. 3.15 Antecedent fuzzy sets

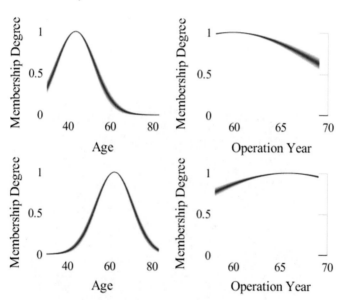

Fig. 3.16 Fuzzy rules

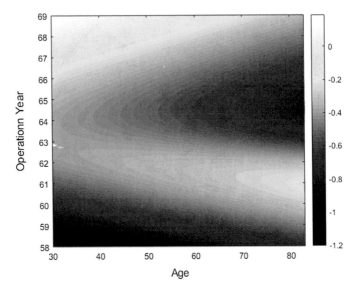

Fig. 3.17 Classification map

As can be noted, the classification is non-linear, these non-linearities are produced by the membership functions and the inference, on the other hand, a map of the uncertainty can be found in Fig. 3.18.

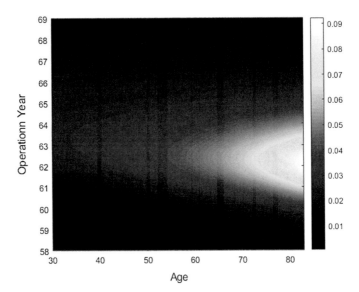

Fig. 3.18 Uncertainty map

The system provides us not only the classification but an uncertainty ratio that can be useful according to the uncertainty observed by the system for a specific set of input data.

3.13   Shadowed Type-2 Fuzzy Diagnosis System with PSO (ST2 + PSO FDS)

This approach is interesting because presents an alternative to reduce the computational cost of the Fuzzy Diagnosis System the alternative of Shadowed Type-2 Fuzzy Systems can be implemented for every GT2 Fuzzy System. We propose this approach that optimizes the Footprint of Uncertainty based on metaheuristics, in this case, the PSO algorithm, obtained by this way an optimal uncertainty based on the training data. For this approach, the input is T1 Gaussian MFs. The flowchart for the design of this approach is illustrated in Fig. 3.19.

The example of the design of a diagnosis system based on the proposed approach are presented as follows.

Starting with the input data observed in Fig. 3.20 (corresponding to Haberman's dataset), the first computation is the estimation of the input membership function parameters, in this case, they are Type-1 Gaussians MFs.

The generated Type-1 Gaussian MFs are obtained by the statistical methodology (Sect. 2.11) and after tuning based on the PSO algorithm (Sect. 2.13) obtaining by this way the GT2 MFs that can be observed in Fig. 3.21.

Based on the kind of rule selector we obtain the following rules (Fig. 3.22).

With the parameters obtained before then is possible to estimate the Sugeno coefficients and obtain a map that explains the classification of the FDS generated.

The classification map can be found in Fig. 3.23.

The other interesting information that we have is the uncertainty map, that illustrates the classification regions with more or less uncertainty estimated by the FDS, this can be observed in Fig. 3.24.

3.14   No-Singleton General Type-2 Fuzzy Diagnosis System (NS GT2 FDS)

The No-Singleton General Type-2 Fuzzy Diagnosis System is similar to the previously introduced General Type-2 Fuzzy Diagnosis System, the difference is in the kind of input data because in this case, the input is non-singleton. For this approach, the input is T1 Gaussian MFs. The flowchart for the design of this system is illustrated in Fig. 3.25.

The example of the design of a diagnosis system based on the proposed approach are presented as follows.

Fig. 3.19 GT2 + PSO FDS
design

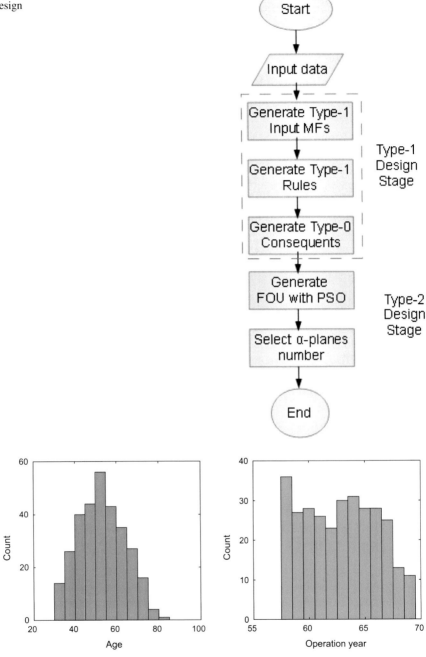

Fig. 3.20 Haberman's features histogram

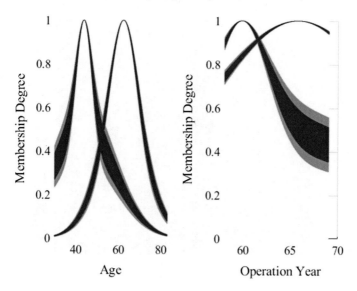

Fig. 3.21 Antecedent fuzzy sets

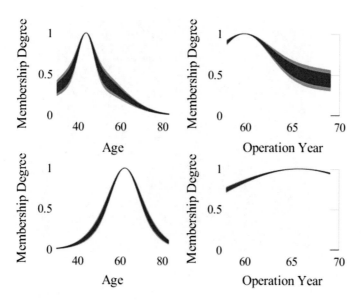

Fig. 3.22 Fuzzy rules

Starting with the input data observed in Fig. 3.26 (corresponding to the Haberman's dataset), and remembering that the first step is the estimation of the parameters of the antecedent MFs.

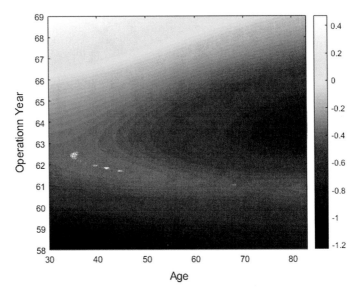

Fig. 3.23 Classification map

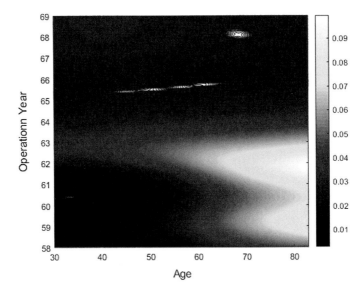

Fig. 3.24 Uncertainty map

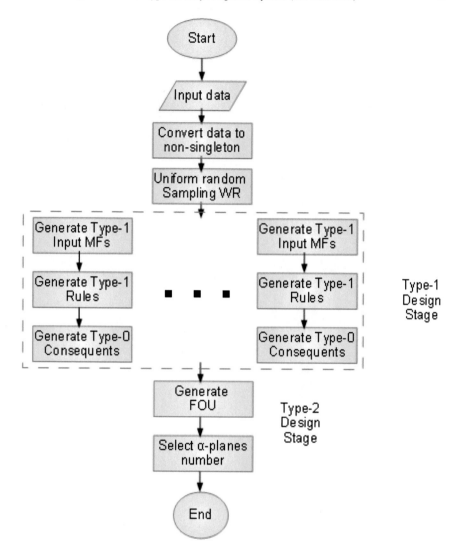

Fig. 3.25 NS GT2 FDS design

Similar to the other approach, the first step is obtaining of the center of the clusters and the membership degree corresponding to each cluster using the FCM algorithm. This step is realized by the methodology proposed in 2.12 by using the principle of embedded T1 MFs for design the GT2 MFs. Figure 3.27 illustrates the multiple T1 MFs and Fig. 3.28 illustrate the generated GT2 MFs.

Based on the generated GT2 MFs the second step is the generation of the fuzzy rules and this step is realized as is explained in 2.21. The rules generated are illustrated in Fig. 3.29.

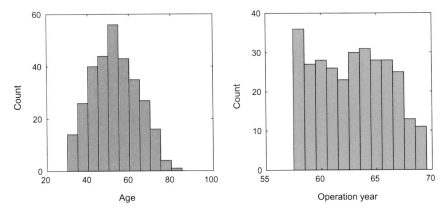

Fig. 3.26 Attributes histogram

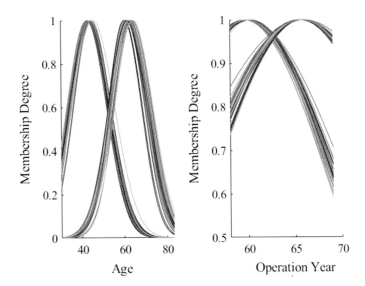

Fig. 3.27 Embedded type-1 MFs

Based on the rules and the fuzzy sets is possible to apply LSE to obtain the Sugeno coefficients as is explained in Sect. 2.3.

The classification map can be found in Fig. 3.30.

As is illustrate in Fig. 3.30, the classification is non-linear, on the other hand, a map of the uncertainty can be found in Fig. 3.31.

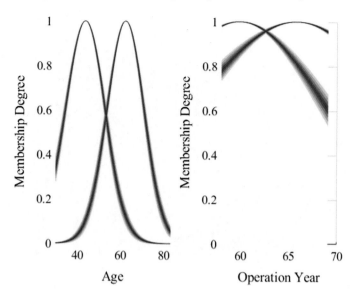

Fig. 3.28   Antecedent fuzzy sets

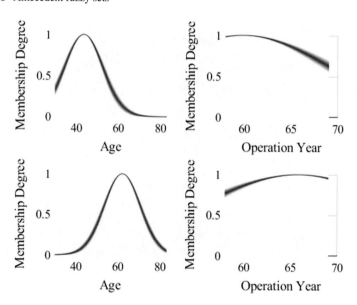

Fig. 3.29   Fuzzy rules

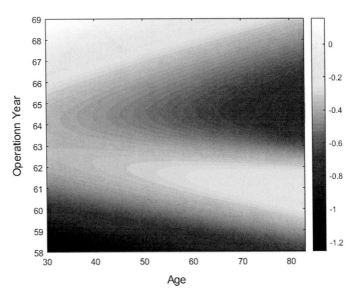

Fig. 3.30 Classification map

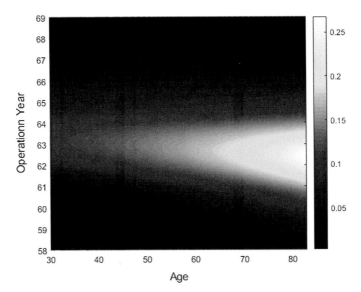

Fig. 3.31 Uncertainty map

3.15   General Type-2 Fuzzy Diagnosis System with SVM (SVM GT2 FDS)

This approach consists in hybridization between GT2 FDS and SVM algorithm, remembering that the SVM algorithm is one of the most used algorithms for classification in the literature, we propose this hybrid approach based on the concept of the hyperplane.

The main idea for this hybridization is the use of the fuzzy rules as kernels of an optimized hyperplane (by SVM algorithm) to obtain a better classification than an SVM classifier or a GT2 FDS. The flowchart for the design of this system can be appreciated in Fig. 3.32.

The same example will be explained next: Considering the input data observed in Fig. 3.33 (corresponding to the Haberman's dataset), and looking for obtaining the parameters of the antecedent MFs.

This step is realized by the steps proposed in Sect. 2.12 and consists of the generation of the FOU using the concept the embedded Type-1 (Fig. 3.34). Figure 3.35 illustrates the obtained GT2 MFs.

The fuzzy rules are the same as the GT2 FDS (Explained in Sect. 2.21), and they are illustrated in Fig. 3.36.

The difference of this approach with respect to the other is in the output because this system is hybridization and the output is a hyperplane optimized by the methodology of SVM. The output equation can be found in (3.2), for a single α-plane, and (3.3) describe the aggregation of the α-planes to obtain the approximated output.

$$\tilde{A}_\alpha = \{((x, u)) | u \in [0, 1], \mu_{\tilde{A}}(x) = \alpha\} \tag{3.2}$$

$$\tilde{Z} = \bigcup \tilde{Z}_\alpha \tag{3.3}$$

The classification map can be found in Fig. 3.37.

As can be observed, the classification is non-linear, on the other hand, a map of the uncertainty can be found in Fig. 3.38.

3.16   General Type-2 Fuzzy Diagnosis System with Asymmetric Uncertainty (ASY GT2 FDS)

This is maybe the most complex approach, considering the antecedent GT2 MFs. The main idea in this approach is obtaining of the parameters of the antecedent GT2 MFs by a combination of the concept of T1 embedded MFs, the concepts of Boxplot, and optimized by the PSO algorithm. The flowchart for the design of this system is illustrated in Fig. 3.39.

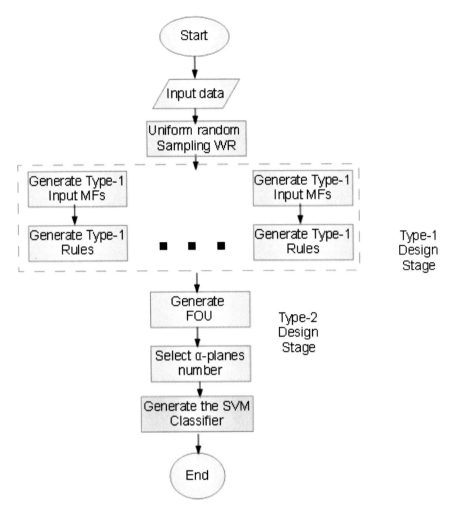

Fig. 3.32 GT2 FDS design

The example of the design of a diagnosis system based on the proposed approach are presented as follows.

Starting with the input data observed in Fig. 3.40 (corresponding to the Haberman's dataset), and remembering that the first step is the estimation of the parameters of the antecedent MFs.

The first step, in this case, is the obtained of the embedded Type-1 MFs, that will be used in the next steps to define the FOU, Fig. 3.41 illustrates the obtained set of T1 MFs.

Based on this set of T1 MFs, the second step is obtaining the points of the boxplot for every point of the primary domain of the T1 MFs, as can be observed in Fig. 3.42.

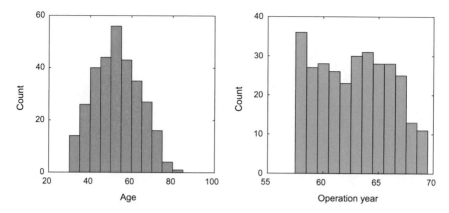

Fig. 3.33 Attributes histogram

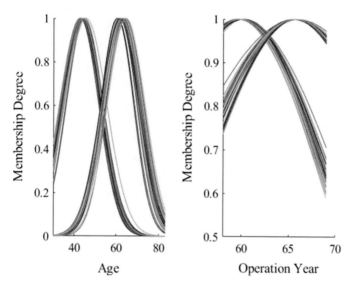

Fig. 3.34 Embedded type-1 MFs

As is detailed in Sect. 3.14, these points can be considered as the FOU and the COU respectively, and offer a non-asymmetric distribution of the uncertainty in the functions. Considering this, the objective is the finding of GT2 MFs that represent a better way the obtained FOU and COU, and this will be realized by an optimization-based on the PSO algorithm, the obtained GT2 MFs can be observed in Fig. 3.43.

The rules for this approach are the same as the GT2 FDS and the NS GT2 FDS and are explained in detail in Sect. 2.2 and appreciated in Fig. 3.44.

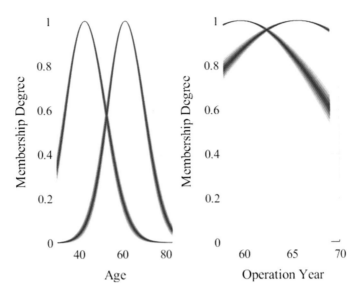

Fig. 3.35 Antecedent fuzzy sets

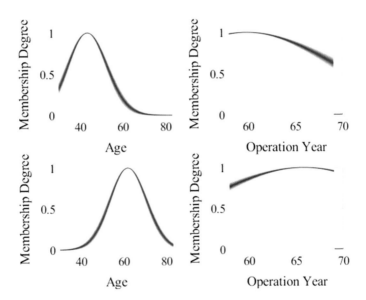

Fig. 3.36 Fuzzy rules

Again, the estimation of the Sugeno Coefficients is realized by the LSE algorithm as is detailed in Sect. 2.3.

The classification map can be found in Fig. 3.45.

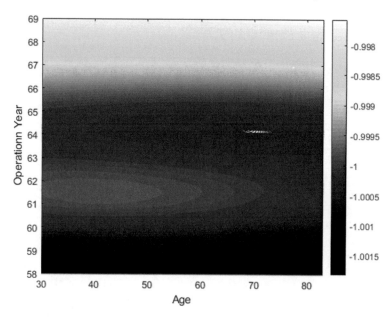

Fig. 3.37 Classification map

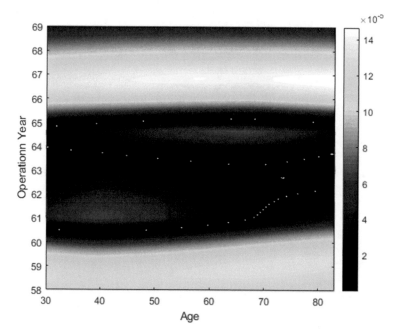

Fig. 3.38 Uncertainty map

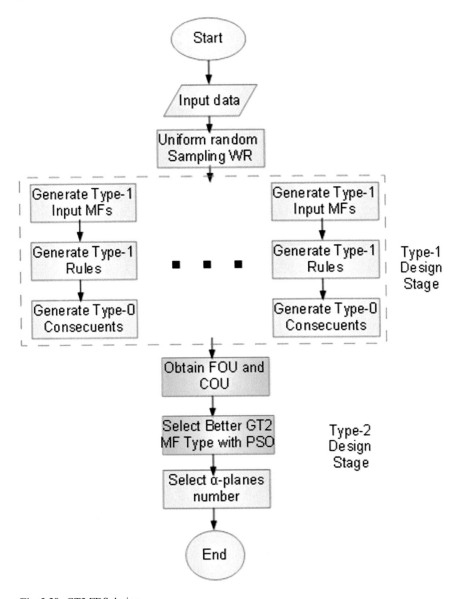

Fig. 3.39 GT2 FDS design

As can noted, the classification is non-linear, on the other hand, a map of the uncertainty can be found in Fig. 3.46.

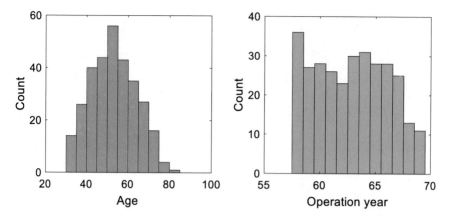

Fig. 3.40   Attributes histogram

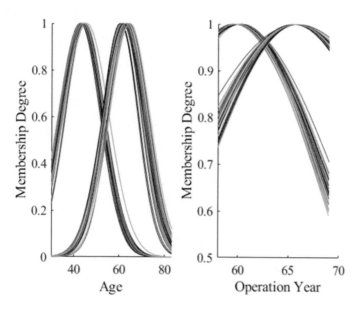

Fig. 3.41   Embedded type-1 MFs

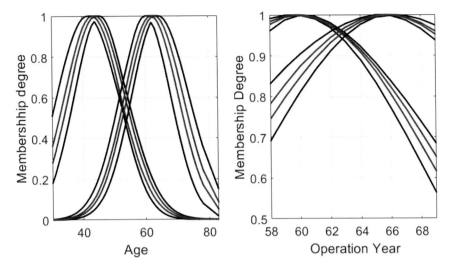

Fig. 3.42 Boundaries

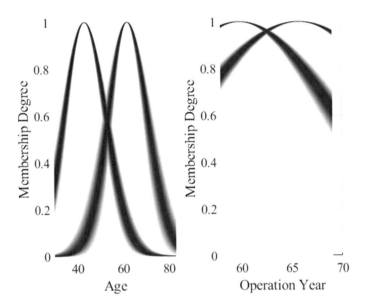

Fig. 3.43 Antecedent fuzzy sets

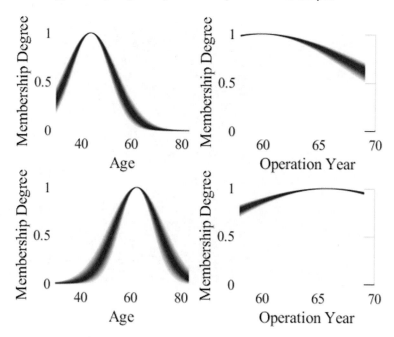

Fig. 3.44 Fuzzy rules

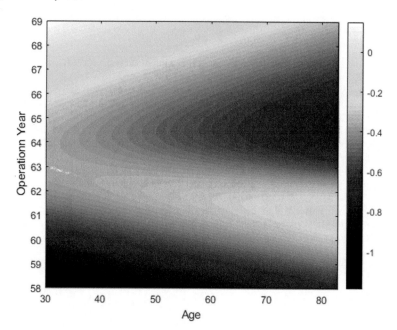

Fig. 3.45 Classification map

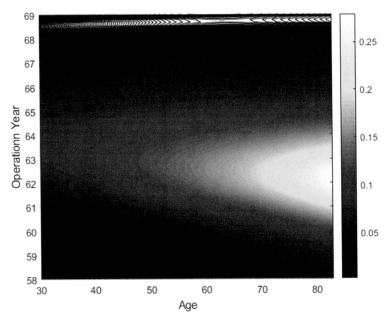

Fig. 3.46 Uncertainty map

References

1. J.-S.R. Jang, ANFIS: adaptive-network-based fuzzy inference system. IEEE Trans. Syst. Man Cybern. **23**(3), 665–685 (1993). https://doi.org/10.1109/21.256541
2. E.-H. Kim, S.-K. Oh, W. Pedrycz, Design of reinforced interval type-2 fuzzy C-means-based fuzzy classifier. IEEE Trans. Fuzzy Syst. **26**(5), 3054–3068 (2018). https://doi.org/10.1109/TFUZZ.2017.2785244
3. E. Ontiveros, P. Melin, O. Castillo, Comparative study of interval type-2 and general type-2 fuzzy systems in medical diagnosis. Inf. Sci. **525**, 37–53 (2020). https://doi.org/10.1016/j.ins.2020.03.059
4. E. Ontiveros-Robles, P. Melin, Toward a development of general type-2 fuzzy classifiers applied in diagnosis problems through embedded type-1 fuzzy classifiers. Soft Comput. **24**(1), 83–99 (2020). https://doi.org/10.1007/s00500-019-04157-2
5. Y. Zeng, Z. Xu, Y. He, Y. Rao, Fuzzy entropy clustering by searching local border points for the analysis of gene expression data. Knowl.-Based Syst. **190**, 105309 (2020). https://doi.org/10.1016/j.knosys.2019.105309
6. P.M. Cincotta, I.I. Shevchenko, Correlations in area preserving maps: a Shannon entropy approach. Phys. Nonlinear Phenom. **402**, 132235 (2020). https://doi.org/10.1016/j.physd.2019.132235

Chapter 4
Experimental Results

In this chapter the results obtained for the previously presented approaches are presented, in order to compare the developed systems based on the performance measured with the accuracy.

The datasets used for experimentation are widely used in the literature to compare the performance of classifiers applied in diagnosis systems. Table 4.1 summarizes the used datasets.

4.1 Performance Comparison Based on Cross-Validation

This section provides the summarize of the performance obtained by the proposed system and other of the most used systems in the realm of decision support and classification (Naïve Bayesian, Decision Tree, and Monolithic ANN). The set up consists of the cross-validation with different values of K (3, 5 and 10) reporting the average of 10 experiments (this set up is selected in order to be compared with the literature results. Tables 4.2, 4.3, and 4.4 document the obtained results for the datasets introduced in Table 4.1 (The yellow highlighted cells note the better performance for the dataset), in order to make a comparison with respect the conventional classifiers.

Based on the results documented in the presented experiments, based on the K-Folds cross-validation, Tables 4.5, 4.6 and 4.7 summarize the results of all datasets, for K = 3, K = 5 and K = 10 respectively, giving 1 point for every victory and using the standard deviation as an undrawn criterion (This because a lower standard deviation means a more precise system).

As can be observed, the proposed approaches obtain better results in most of the cases, especially the hybrid approach of GT2 with SVM, and the GT2 with non-singleton inputs, of course, they are only four alternatives and based on the proposed approaches and methodologies can be proposed alternatives that provide better performance.

Table 4.1 Benchmark datasets (from UCI repository)

Dataset name	Attributes	Abbreviation	Instances	Data entropy
Breast cancer Wisconsin (original) data set	9	BCW	699	0.92932
Haberman's Survival Data Set	3	Haber	306	0.83376
Fertility data set	10	Fert	100	0.52936
Indian liver data set	9	Indian	583	0.86409
Breast cancer Wisconsin (diagnostic) data set	32	BCWD	569	0.95264
Pima Indians diabetes data set	8	Pima	768	0.93313
Statlog (heart) data set	13	Heart	270	0.99108
Mammographic mass data set	5	MMass	825	0.99934
Immunotherapy data set	8	Inmu	90	0.7436
Cryotherapy data set	7	Cryo	90	0.99679
Breast cancer Wisconsin (prognostic)	34	WPDC	198	0.79065
Breast cancer coimbra	10	Coimbra	116	0.99227
SPECT heart	22	Spect	267	0.73375

Some examples of performances obtained for Fuzzy Systems in the literature can be noticed in Tables 4.8, 4.9 and 4.10 for $K = 3$, $K = 5$ and $K = 10$ respectively.

As can be noticed, the performed results in Sect. 4.4 are very competitive with respect to the reference of the literature, and even when do not have enough information for a statistical test, the proposed approach provides with an interesting alternative for diagnosis in comparison with the fuzzy approaches of the literature.

4.2 Performance Comparison Based on Z-Test

This section provides a statistical comparison of the proposed approaches in comparison with a conventional Type-1 FDS. The details of the statistical test can be found in Table 4.11.

The experiments consist of thirty experiments performed with the methodology of hold-out validation but compared with the Z-Test the main idea is the observation of the improvement of the use of Type-2 for diagnosis and measured with a statistical test (Tables 4.12, 4.13, 4.14 and 4.15 for ST2 PSO, GT2 ASY, GT2 SVM and GT2 NS respectively) where the green highlight corresponds with a passed test and the yellow highlight means that the evidence is not enough to pass the test.

As can be observed in Table 4.12, the evidence is enough to say that the Type-2 Approach is better than Type-1 in six datasets of thirteen, in the other cases the evidence is not enough but in every case, the rank of the sum is higher for the Type-2

Table 4.2 K = 3 cross validation

Dat.		ST2 PSO	GT2-ASY	GT2-SVM	GT2-NS	DT	NV	ANN
BCW	Mean	0.96552	0.96452	0.96366	0.96552	0.94263	0.95995	0.96194
	S.D.	0.00274	0.00126	0.001838	0.00101	0.0051	0.0027	0.00509
Haber	Mean	0.73791	0.7415	0.73431	0.74085	0.67614	0.74444	0.71275
	S.D.	0.01665	0.01235	0.004858	0.01044	0.01637	0.00811	0.03587
Fert	Mean	0.79564	0.8303	0.87778	0.84949	0.80909	0.8101	0.80707
	S.D.	0.02764	0.0309	0.00303	0.02279	0.04027	0.03636	0.0479
Indian	Mean	0.69966	0.70957	0.53599	0.66922	0.6566	0.56067	0.69948
	S.D.	0.00945	0.01528	0.041951	0.02321	0.01862	0.00426	0.0072
BCWD	Mean	0.95732	0.95344	0.93298	0.95485	0.93228	0.93792	0.95697
	S.D.	0.00293	0.00317	0.003346	0.00317	0.00875	0.00408	0.00542
Pima	Mean	0.7612	0.76289	0.76823	0.76367	0.69844	0.75677	0.75456
	S.D.	0.00531	0.00557	0.006272	0.00854	0.0141	0.0057	0.01293
Heart	Mean	0.81593	0.81296	0.83185	0.80222	0.72593	0.84111	0.78
	S.D.	0.01022	0.01418	0.013578	0.01436	0.01844	0.00607	0.0243
Coimbra	Mean	0.66452	0.68351	0.71894	0.68011	0.68838	0.61567	0.66893
	S.D.	0.03024	0.03444	0.023474	0.02108	0.02822	0.02214	0.04599
Cryo	Mean	0.84778	0.83556	0.86	0.84444	0.85222	0.84	0.79222
	S.D.	0.02437	0.03712	0.031505	0.02767	0.02984	0.01805	0.04995
MMass	Mean	0.83503	0.83211	0.82946	0.84145	0.8046	0.82496	0.82545
	S.D.	0.00755	0.01698	0.002814	0.00443	0.01091	0.00345	0.00854
WPDC	Mean	0.71364	0.68737	0.74394	0.72222	0.7	0.66667	0.73283
	S.D.	0.02495	0.03481	0.029712	0.02166	0.03099	0.0166	0.03158
Spect	Mean	0.70545	0.62846	0.75543	0.66816	0.74232	0.68801	0.74757
	S.D.	0.03621	0.02506	0.022969	0.03143	0.02374	0.00903	0.0347
Inmu	Mean	0.73667	0.53333	0.8	0.70111	0.8	0.78	0.69444
	S.D.	0.02225	0.03583	0.02166	0.05269	0.02676	0.03435	0.05194

approach, this means that maybe the data sets do not have enough noise and the Type-1 system offer a very good solution.

As can be noticed in Table 4.13, the results are similar than Table 4.12, the results obtained by the Type-2 GT2 with asymmetric uncertainty are better in seven datasets, the reason can be related with the good result obtained by both approaches and in the difficult of the datasets, the GT2 system is better in the most difficult datasets.

In Table 4.14 can be observed a similar tendency, the improvement exists in eight cases but in the rest of the cases is not passed the test, the obtained results are good for this approach compared with respect the other proposed approaches.

Finally, in Table 4.15 we have that the test is passed only for five datasets, the results are not good because the evidence is not significate in most of the datasets.

Table 4.3 K = 5 cross validation

Dat.		ST2 PSO	GT2-ASY	GT2-SVM	GT2-NS	DT	NV	ANN
BCW	Mean	0.9629	0.9637	0.9637	0.9660	0.9419	0.9598	0.9601
	S.D.	0.0050	0.0009	0.0024	0.0016	0.0035	0.0008	0.0065
Haber	Mean	0.7443	0.7423	0.7282	0.7433	0.6875	0.7505	0.7298
	S.D.	0.0097	0.0069	0.0058	0.0069	0.0285	0.0040	0.0254
Fert	Mean	0.8131	0.8520	0.8800	0.8720	0.8040	0.8190	0.8220
	S.D.	0.0230	0.0160	0.0000	0.0133	0.0323	0.0330	0.0117
Indian	Mean	0.7027	0.7134	0.5259	0.6937	0.6564	0.5570	0.7036
	S.D.	0.0123	0.0079	0.0304	0.0141	0.0214	0.0037	0.0088
BCWD	Mean	0.9614	0.9529	0.9356	0.9545	0.9379	0.9391	0.9605
	S.D.	0.0048	0.0029	0.0020	0.0033	0.0074	0.0023	0.0048
Pima	Mean	0.7641	0.7625	0.7664	0.7631	0.7037	0.7558	0.7590
	S.D.	0.0049	0.0051	0.0039	0.0048	0.0111	0.0054	0.0106
Heart	Mean	0.8196	0.8170	0.8352	0.8293	0.7404	0.8437	0.7778
	S.D.	0.0083	0.0083	0.0098	0.0084	0.0246	0.0068	0.0168
Coimbra	Mean	0.6988	0.6909	0.7406	0.6866	0.6926	0.6229	0.6780
	S.D.	0.0243	0.0196	0.0262	0.0193	0.0325	0.0223	0.0276
Cryo	Mean	0.8572	0.8693	0.8734	0.8629	0.8670	0.8417	0.8344
	S.D.	0.0190	0.0186	0.0152	0.0142	0.0292	0.0115	0.0440
MMass	Mean	0.8405	0.8312	0.8302	0.8439	0.8032	0.8267	0.8293
	S.D.	0.0046	0.0180	0.0004	0.0039	0.0054	0.0029	0.0089
WPDC	Mean	0.7369	0.7221	0.7303	0.7528	0.6872	0.6672	0.7200
	S.D.	0.0249	0.0282	0.0226	0.0206	0.0227	0.0120	0.0361
Spect	Mean	0.7236	0.6509	0.7755	0.6800	0.7551	0.6887	0.7634
	S.D.	0.0356	0.0282	0.0265	0.0200	0.0281	0.0076	0.0194
Inmu	Mean	0.7511	0.5567	0.7922	0.7000	0.8078	0.7744	0.7156
	S.D.	0.0187	0.0300	0.0258	0.0425	0.0286	0.0337	0.0529

As can be observed, the statistical test gives us an interesting conclusion some of the proposed approaches do not show improvement in comparison with respect Type-1, of course, this does not mean that the approaches are not interesting because provide us with the parameter of uncertainty, but is necessary to select the FDS considering the application environment and the uncertainty sources in our problem.

Table 4.4 K = 10 cross validation

Dat.		ST2 PSO	GT2-ASY	GT2-SVM	GT2-NS	DT	NV	ANN
BCW	Mean	0.9641	0.9634	0.9657	0.9644	0.9411	0.9599	0.9635
	S.D.	0.0030	0.0007	0.0025	0.0004	0.0037	0.0006	0.0035
Haber	Mean.	0.7377	0.7407	0.7250	0.7413	0.6810	0.7500	0.7230
	S.D.	0.0063	0.0049	0.0056	0.0043	0.0173	0.0045	0.0139
Fert	Mean	0.8343	0.8540	0.8800	0.8680	0.8100	0.8350	0.8200
	S.D.	0.0196	0.0066	0.0000	0.0060	0.0313	0.0150	0.0297
Indian	Mean	0.7060	0.7163	0.5280	0.6499	0.6444	0.5579	0.7012
	S.D.	0.0064	0.0068	0.0155	0.0117	0.0137	0.0029	0.0118
BCWD	Mean	0.9575	0.9504	0.9386	0.9520	0.9398	0.9396	0.9613
	S.D.	0.0050	0.0025	0.0016	0.0020	0.0036	0.0017	0.0051
Pima	Mean	0.7645	0.7611	0.7632	0.7593	0.6922	0.7558	0.7546
	S.D.	0.0061	0.0031	0.0070	0.0049	0.0108	0.0046	0.0052
Heart	Mean	0.8237	0.8215	0.8300	0.8256	0.7670	0.8426	0.7778
	S.D.	0.0080	0.0104	0.0077	0.0087	0.0157	0.0058	0.0230
Coimbra	Mean	0.7114	0.7139	0.7449	0.7055	0.7220	0.6243	0.6886
	S.D.	0.0134	0.0229	0.0165	0.0178	0.0311	0.0113	0.0477
Cryo	Mean	0.8696	0.8861	0.8676	0.8714	0.8679	0.8528	0.8319
	S.D.	0.0142	0.0179	0.0173	0.0106	0.0190	0.0071	0.0379
MMass	Mean	0.8420	0.8388	0.8303	0.8444	0.8027	0.8266	0.8287
	S.D.	0.0043	0.0025	0.0001	0.0030	0.0046	0.0018	0.0066
WPDC	Mean	0.7542	0.7358	0.7142	0.7632	0.6837	0.6653	0.7374
	S.D.	0.0217	0.0117	0.0272	0.0158	0.0290	0.0127	0.0375
Spect	Mean	0.7516	0.6727	0.7869	0.7092	0.7554	0.6881	0.7727
	S.D.	0.0162	0.0225	0.0090	0.0128	0.0258	0.0074	0.0236
Inmu	Mean	0.7389	0.5078	0.8167	0.7244	0.8011	0.7878	0.7200
	S.D.	0.0159	0.0410	0.0234	0.0204	0.0265	0.0136	0.0371

Table 4.5 K = 3 results summarize

System	Points
ST2 PSO	1
GT2 ASY	1
GT2 SVM	7
GT2 NS	2
DT	
NV	2
ANN	

Table 4.6 K = 5 results
summarize

System	Points
ST2 PSO	1
GT2 ASY	1
GT2 SVM	4
GT2 NS	3
DT	1
NV	2
ANN	

Table 4.7 K = 10 results
summarize

System	Points
ST2 PSO	1
GT2 ASY	2
GT2 SVM	5
GT2 NS	2
DT	0
NV	2
ANN	1

Table 4.8 3-folds CV

Dataset	LFA [1]
WBCD	97.28
PIMA	78.39
Heart	85.15
Indian	71.36
Haber	75.49

Table 4.9 5-folds CV

Dataset	FCM-G [2]	FRF [3]
AUS	86.19	
WBCD	96.64	97.300
PIMA	75.13	76.530
WDBC		
Heart		82.870
Fert		80.5
Haber	74.2	

Table 4.10 10-folds CV

Dataset	FCM-TSK [4]	B-TSK [5]
AUS	84.59	86.88
WBCD	95.37	97.210
PIMA	76.02	77.22
WDBC	94.46	96.00
Heart	84.45	85.62
Indian	65.48	66.71
Haber	72.92	75.87

Table 4.11 Test parameters

Parameter	
Significance	95%
p	0.05
Ha	$\mu_1 > \mu_2$
Ho	$\mu_1 <= \mu_2$
Critical value	1.645

Table 4.12 ST2 PSO versus type-1

Dataset	T2 Mean	T2 S.D.	T1 Mean	T1 S.D.	Z
Hab	75.1086957	3.70576196	72.826087	3.23662485	3.86277784
Fert	81.183908	6.57113144	83.8888889	7.20510793	-2.05628984
Pima	76.3188406	2.41485983	75.5362319	2.91214205	1.47194893
WDBC	96.1208577	1.46769516	95.0877193	1.25968767	4.49217071
WPDC	73.3898305	5.38029929	70.5084746	6.08912704	2.59180606
Inm	73.5802469	6.60695712	73.0864198	7.58631679	0.35653702
spect	70.1343909	4.74503765	73.7916667	4.60393099	-4.35100443
Mma	84.0080972	2.01278783	83.9676113	2.11303455	0.10494387
WBDC	96.3492063	1.11563704	95.9365079	1.06289962	2.12667523
Heart	82.345679	3.20591715	81.0288066	2.71386619	2.65776086
Coimbra	68.952381	6.11603595	66.5714286	6.30947889	2.06689229
Indian	70.1333333	2.95796707	70.8571429	2.06559112	-1.91928983
Cryo	83.2098765	4.13902583	82.962963	5.46008503	0.2476887

4.3 Computational Cost Comparison

In this section, the proposed systems are compared in terms of computational effort, first based in the computation time.

Table 4.13 GT2 ASY versus type-1

Dataset	T2 Mean	T2 S.D.	T1 Mean	T1 S.D.	Z
Hab	74.3115942	3.99356892	72.826087	3.23662485	2.51387129
Fert	86.1111111	5.77884606	83.8888889	7.20510793	1.68930327
Pima	76.9275362	2.4999895	75.5362319	2.91214205	2.61679809
WDBC	95.2436647	1.46185834	95.0877193	1.25968767	0.6780635
WPDC	73.4463277	4.65887641	70.5084746	6.08912704	2.64262579
Inm	65.9259259	12.4465019	73.0864198	7.58631679	-5.16978674
spect	64.75	4.78060317	73.7916667	4.60393099	-10.7567312
Mma	83.3603239	2.1886077	83.9676113	2.11303455	-1.57415805
WBDC	96.1111111	1.21145536	95.9365079	1.06289962	0.89974721
Heart	81.4814815	4.19268663	81.0288066	2.71386619	0.9136053
Coimbra	69.4285714	6.76726816	66.5714286	6.30947889	2.48027075
Indian	72.1714286	2.67271423	70.8571429	2.06559112	3.48502628
Cryo	85.8024691	6.8348694	82.962963	5.46008503	2.84842008

Table 4.14 GT2 SVM versus type-1

Dataset	T2 Mean	T2 S.D.	T1 Mean	T1 S.D.	Z
Hab	73.6594203	3.56861067	72.826087	3.23662485	1.41022048
Fert	89.1111111	6.14234443	83.8888889	7.20510793	3.96986269
Pima	76.9275362	2.49241656	75.5362319	2.91214205	2.61679809
WDBC	93.0409357	1.68376375	95.0877193	1.25968767	-8.89958349
WPDC	75.1977401	8.28584574	70.5084746	6.08912704	4.21803732
Inm	79.5061728	7.92819474	73.0864198	7.58631679	4.63498121
spect	76.2083333	4.08864451	73.7916667	4.60393099	2.875071
Mma	83.1174089	2.03008617	83.9676113	2.11303455	-2.20382127
WBDC	95.968254	1.09097399	95.9365079	1.06289962	0.1635904
Heart	83.6213992	3.99325415	81.0288066	2.71386619	5.2324667
Coimbra	72.4761905	6.1934059	66.5714286	6.30947889	5.12589288
Indian	71.9047619	1.91284001	70.8571429	2.06559112	2.7779195
Cryo	86.7901235	5.21011058	82.962963	5.46008503	3.83917489

4.4 Computational Cost

The first measure for comparing the computational cost of the proposed systems is the computation time, considering this, Fig. 4.1 illustrates the computation time of the proposed architecture. The results reported are average the time in seconds of the training and testing of the proposed benchmark problems.

Table 4.15 GT2 NS versus type-1

Dataset	T2 Mean	T2 S.D.	T1 Mean	T1 S.D.	Z
Hab	74.1666667	4.17862179	72.826087	3.23662485	2.26861556
Fert	85.5555556	5.46594394	83.8888889	7.20510793	1.26697745
Pima	76.5797101	1.75978078	75.5362319	2.91214205	1.96259857
WDBC	94.9902534	1.15668645	95.0877193	1.25968767	-0.42378969
WPDC	75.0282486	5.80465516	70.5084746	6.08912704	4.06557814
Inm	65.6790123	10.9424789	73.0864198	7.58631679	-5.34805825
spect	67.5	5.75543222	73.7916667	4.60393099	-7.48509864
Mma	84.3724696	1.72591873	83.9676113	2.11303455	1.0494387
WBDC	96.6190476	0.93919442	95.9365079	1.06289962	3.51719365
Heart	81.6460905	3.47340705	81.0288066	2.71386619	1.2458254
Coimbra	69.8095238	7.25374249	66.5714286	6.30947889	2.81097352
Indian	69.0579979	3.54884246	70.8571429	2.06559112	-4.77070354
Cryo	84.1975309	6.62078404	82.962963	5.46008503	1.23844351

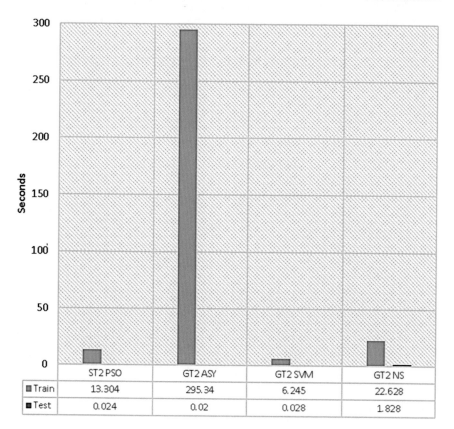

	ST2 PSO	GT2 ASY	GT2 SVM	GT2 NS
■ Train	13.304	295.34	6.245	22.628
■ Test	0.024	0.02	0.028	1.828

Fig. 4.1 Computation time comparison (seconds)

Is interesting that the GT2 with Asymmetric approach demands the higher time for the training step but in the test takes the lower time, the problem with some applications is focused in reduce the computation time in the practice (test time) but in some scenarios the training time cannot be unlimited or very high because exist systems that can be adapted in real-time.

References

1. M. Pota, M. Esposito, G. De Pietro, Likelihood-fuzzy analysis: from data, through statistics, to interpretable fuzzy classifiers. Int. J. Approx. Reason. **93**, 88–102 (2018). https://doi.org/10.1016/j.ijar.2017.10.022
2. C. Fu, W. Lu, W. Pedrycz, J. Yang, Fuzzy granular classification based on the principle of justifiable granularity. Knowl.-Based Syst. **170**, 89–101 (2019). https://doi.org/10.1016/j.knosys.2019.02.001
3. P. Bonissone, J.M. Cadenas, M. Carmen Garrido, R. Andrés Díaz-Valladares, A fuzzy random forest. Int. J. Approx. Reason. **51**(7), 729–747 (2010). https://doi.org/10.1016/j.ijar.2010.02.003
4. E.-H. Kim, S.-K. Oh, W. Pedrycz, Design of reinforced interval type-2 fuzzy C-means-based fuzzy classifier. IEEE Trans. Fuzzy Syst. **26**(5), 3054–3068 (2018). https://doi.org/10.1109/TFUZZ.2017.2785244
5. X. Gu, F.-L. Chung, S. Wang, Bayesian Takagi–Sugeno–Kang fuzzy classifier. IEEE Trans. Fuzzy Syst. **25**(6), 1655–1671 (2017). https://doi.org/10.1109/TFUZZ.2016.2617377

Chapter 5
Discussion of Results

As can be observed in the experimental results, some specific T2 FDS proposed with our methodology demonstrated to obtain better results than three of the most used classifiers (DT, NV, and monolithic ANN) compared with cross-validation with K = 3, 5 and 10, and demonstrate to be competitive with other alternatives provided by the literature and based in Fuzzy Logic too. Also, we probe by a statistical test that our methodology can provide us with T2 FDS that overcome systems based in Type-1 Fuzzy Logic as the ANFIS system (not every T2 FDS demonstrates to be better but they can be improved with elements of the bests T2 FDSs).

Based on the results obtained and the experimentation performed in the development of the proposed approaches, it is possible to compare the different approaches based on the computational effort and performance. Some of these conclusions are listed as follows:

- Based on the input type the conclusions are that the non-singleton inputs require a higher computational cost and a better performance, and this conclusion can be observed in Fig. 5.1.

- Based in the approximation of the Generalized Type-2 we explore three alternatives, the conventional α-planes representation, the variation of High-Order α-planes that introduce the Newton Cotes quadrature to reduce computational cost and the Shadowed Type-2 Fuzzy Logic, the conclusion is that all variations obtain almost the same performances but the Shadowed Type-2 Fuzzy Logic is the approach that requires the less computational effort, the High Order α-planes is very similar in its requirements and finally, the conventional α-planes approximation requires the high computational effort. These conclusions can be noticed in Fig. 5.2.

- In the rule's generation was explored three variations, the Fuzzy Entropy (similar than the process used in the C4.5 tree with the Shannon Entropy), Descendent Gradient and the rules of the first ANFIS proposed by Jang. Comparing these

© The Author(s), under exclusive license to Springer Nature Switzerland AG 2021   73
P. Melin et al., *New Medical Diagnosis Models Based on Generalized Type-2 Fuzzy Logic*,
SpringerBriefs in Computational Intelligence,
https://doi.org/10.1007/978-3-030-75097-8_5

Fig. 5.1 Comparison based
on the input type

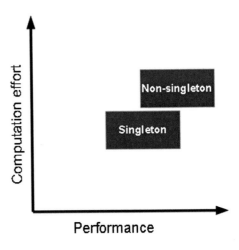

Fig. 5.2 Comparison based
in the type-2 approximation

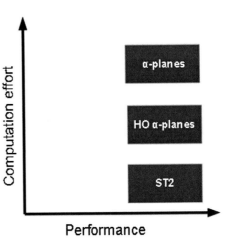

approaches, we found that the ANFIS rules obtain the better performance and require the lower computational cost, obtain systems more robust. On the other hand, the Fuzzy Entropy and Descendent Gradient are affected for the overfitting, for classifier cannot obtain better results but maybe can be applied in curves fitting. Figure 5.3 illustrates these conclusions.

- Finally, based on the FOU estimation was explored different variations, first the concept of Embedded Type-1 but also a variation for asymmetric uncertainty, and the implementation of metaheuristics, in this case, the PSO algorithm. The best performance, in this case, is obtained with the asymmetric uncertainty, this can be related with the good modeling of the uncertainty, but requires the higher computational effort, on the other hand, the symmetric uncertainty obtained with the concept of embedded type-1 obtain good performance and lower computational

Fig. 5.3 Comparison based on the rules generation

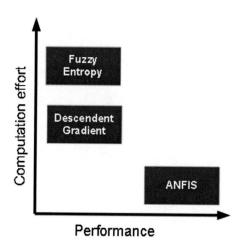

Fig. 5.4 Comparison based on the uncertainty estimation

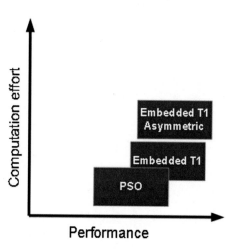

cost than its alternative asymmetric, and finally, the optimized with metaheuristics requires the lower computational cost but obtain the worst performances. These conclusions are illustrated in Fig. 5.4.

Chapter 6
Conclusions

In the present work, it was proposed a methodology for the automatic design of Fuzzy Diagnosis Systems (based in Generalized Type-2 Fuzzy Logic) this methodology consists of different steps that perform specific tasks but can be realized by different methods (have a modular structure where every module can be substituted or improved by separated).

The main idea in the work is the improvement of the conventional fuzzy diagnosis systems (for example ANFIS) with the inclusion of the Type-2 with its uncertainty handling, some of the focus of the paper is in the generation of the Footprint of Uncertainty based in the data (in many works the FOU is arbitrary or based in by expert experience), in this point was proposed different alternatives for obtaining of the FOU, based in statistical measurements, in concepts as the embedded Type-1 fuzzy membership functions and variations, and also the optimization-based in metaheuristics.

Other of the goals of the work is the reduction of the computational effort of the proposed approach, in this case, was explored alternatives as α-planes representation, Shadowed Type-2 approximation, and was proposed the α-planes with high-order aggregation that obtains good results.

Some of the contributions of the work are the multiple variations for the modelling of the FOU in General Type-2 FDSs, in addition with the study that give us evidence of the relevance of this parameter in the performance of Fuzzy Classifiers, is very clear than the selection of the FOU cannot be arbitrary and in the present thesis was proposed alternatives to obtain a justifiable FOU in the systems. On the other hand, was explored alternatives for the rules selection obtaining that in some cases is desirable a robust and scalable system than a specialized system, this is because in many cases, the training data do not have enough relation with the testing data and an overtraining can produce many problems in the real performance of the system.

An interesting contribution of the present thesis is the knowledge generated and that allow to have a criterion for the design of the Fuzzy Diagnosis Systems, the results

P. Melin et al., *New Medical Diagnosis Models Based on Generalized Type-2 Fuzzy Logic*, SpringerBriefs in Computational Intelligence, https://doi.org/10.1007/978-3-030-75097-8_6

obtained that generate a discussion in Chap. 5 and provide us with an overview about how many affect the design decisions in this realm (diagnosis problems.).

As future works is interesting to build a system that select the attributes of the FDS (FOU obtaining, rules selection, type-2 modelling, e.g.) based in the data features, in order to obtain the better FDS for a specific problem. Also explore more hybrid approaches with modular architectures or hierarchical approaches.

Printed in the United States
by Baker & Taylor Publisher Services